Science and Environment in Chine

Urban and Industrial Environments

Series editor: Robert Gottlieb, Henry R. Luce Professor of Urban and Environmental Policy, Occidental College

Science and Environment in Chile

The Politics of Expert Advice in a Neoliberal Democracy

Javiera Barandiarán

The MIT Press
Cambridge, Massachusetts
London, England

This book was set in Stone Serif by Westchester Publishing Services. Printed and bound in the United States of America.

Library of Congress Cataloging-in-Publication Data

Names: Barandiaran, Javiera, author.
Title: Science and environment in Chile : the politics of expert advice in a neoliberal democracy / Javiera Barandiaran.
Description: Cambridge, MA : MIT Press, [2018] | Series: Urban and industrial environments | Includes bibliographical references and index.
Identifiers: LCCN 2017053776 | ISBN 9780262038201 (hardcover : alk. paper) | ISBN 9780262535632 (pbk. : alk. paper)
Subjects: LCSH: Environmental policy--Chile. | Environmental policy--Economic aspects--Chile. | Science and state--Chile. | Neoliberalism--Chile. | Chile--Politics and government--1988–
Classification: LCC GE190.C5 B37 2018 | DDC 333.70983--dc23 LC record available at https://lccn.loc.gov/2017053776

10 9 8 7 6 5 4 3 2 1

Contents

Preface

In the two years before this book appeared in print, political and intellectual leaders were particularly fretful about the quality of democracy. Outgoing US president Barack Obama, speaking shortly after the November 2016 election, warned global society that "if everything seems to be the same and no distinctions are made [between truth and fiction], then we won't know what to protect. We won't know what to fight for. And we can lose so much of what we've gained in terms of … democratic freedoms."[1] Some months earlier, after another momentous election—that for the United Kingdom to leave the European Union—scholar Andy Stirling wrote that the new normal seemed to be that "as truths are neglected, democracies weaken in their capacities to build truths." He was reacting to an election campaign in which lies had become routinized and narrow elites had come to occupy democratic institutions, resulting in "corrosive exclusion, cynicism, fear and pent-up violence."[2] Some events over the next year would lend credence to these fears, as social divisions deepened and acts of terrorism multiplied—consider, for instance, white supremacist rallies such as that held in Charlottesville, Virginia. In the United States, moreover, the new administration of Donald J. Trump quickly moved to deny the science behind climate change and championed instead the currency of "alternative facts."

The experience of Chile recounted in this book provides a context for reflecting on these shifts in the value of truth in democracy. One driver of this shift that this book illuminates is neoliberalism, a political ideology that privileges market mechanisms and private initiative over the actions of state agencies, including those that participate in knowledge production. Neoliberalism is by no means the only relevant driver, and national differences are of utmost importance. Nevertheless, when looking across societies for global trends, it is one that stands out. This book focuses particularly on what neoliberalism has meant for the production, credibility, and public usefulness of scientific knowledge in Chile. For all the scholarly and policy writing on Chile's experiments with neoliberal policies and institutions, knowledge production has received relatively little attention. Yet as this book argues, knowledge production has enormous potential to transform how the state operates and conceives of itself. Moreover, as the following list shows, many actors have found the "neoliberalization" of knowledge production to be highly profitable:

1. Higher education: Scholars have for years been warning against the inequities and inefficiencies that result from privatizing universities (Kirp 2003; Geiger 2004; Krimsky 2003; Mirowski 2011; Slaughter and Rhoades 2004). Trump's own "Trump University," which closed after multiple lawsuits for fraud, is a case in point. Chile presents a cautionary tale about privatized universities. In the 1990s, several officials transitioned from the military government to the position of university CEO. They invented a scheme to channel publicly subsidized student loans into mortgage payments, making higher education an accessory to real estate interests. The rich got richer, students gained unmanageable amounts of debt, and many universities became socially conservative (Monckeberg 2007). In 2011, the system imploded as students occupied the streets demanding free and quality education (Figueroa 2013). They forced several especially substandard universities to close, and in 2016, Congress approved free tuition for the poorest 50 percent of incoming students.

2. K–12 education: President Trump's education secretary, Betsy DeVos, has long championed "school choice" through vouchers. This scheme channels public money to privately run schools on the assumption that schools will compete for parents' preferences, thereby improving their performance. Yet as the *Washington Post* reminded US readers, "A large-scale, countrywide experiment in school vouchers has taken place—in Chile. And Chile offers an instructive and cautionary tale about how school vouchers affect education" (published January 17, 2016). Vouchers in Chile improved education for some, but entrenched inequalities that have hurt working families. This time the system imploded nearly thirty years after it was adopted: since 2006, public school students have

gone on strike and occupied their school buildings, demanding better quality and less segregated education. But business interests as well as middle-class parents anxious to protect their fragile socioeconomic status have blocked these efforts. Reforms approved in 2015 have not been enough to turn around this system that disadvantages poor children.

3. Science, the topic of this book: In the United States, several cases have been documented of scientists, while working for corporate interests, manipulating or fabricating results to suit their funders' interests. Sometimes these examples are egregious (e.g., Oreskes and Conway 2011). Other times they are more subtle, as in a piece by the *New York Times* denouncing think tanks like the Brookings Institution for "amplifying Corporate America's influence" on policy makers while doing ostensibly "neutral" science (published August 7, 2016).

These are not isolated cases or policies. Rather, each of the above is an instance of a shift in the logic of public actions and responsibilities—with negative consequences for the state's ability to regulate, among other things, the quality of the environment (Lave 2012). Focusing on Chile, this book details the consequences for environmental governance when state agencies lack the capacity to produce an authoritative body of scientific knowledge. Among other practices, an archetypal neoliberal state would pursue policies not because anyone has any reason to believe they will produce their stated goals—whether to reduce pollution or balance the budget—but rather because they are expedient for the state and profitable for someone in the private sector. For advocates of neoliberal principles, this is desirable; in their view, if private actors are moved by profit, they are being honest. Public actors, by contrast, have complex motivations that can appear duplicitous and ineffective. Better, they argue, for the state to concentrate on ensuring that everyone plays according to the same rules and procedures.

I began this project wanting to understand why public dissatisfaction in Chile seemed so high, despite the ostensible success of the transition to democracy. Chile has changed dramatically since I was born there close to the midpoint of the military dictatorship. Ending decades of political violence, the transition itself was peaceful and the resulting democracy has been stable. Total and per capita wealth have grown dramatically as has access to consumer goods. Yet trust in public institutions continues to fall and cynicism seems rife (Garcés 2012). As a graduate student in the United States in 2008–2009, however, I was having difficulty convincing my professors that Chilean universities were interesting to study and seeing how to apply the theories I was learning to explain the seemingly yawning gap between the myth of Chile's success and lived realities of many Chileans I knew.

Since I began working on this project, Chile continued to change dramatically, as did my chosen field: science and technology studies (STS). In 2014, the Society for the Social Studies of Science (4S) held its annual meeting in Buenos Aires, Argentina, as part of an effort to "go global." An English-language edited volume focused on science in Latin America was published (Medina, Marques, and Holmes 2014), numerous Latin American scholars have assumed leadership positions within 4S, and, in 2017, 4S awarded Argentine Venezuelan scholar Hebe Vessuri the John Desmond Bernal Prize for her distinguished contributions to the field. Interest in studies of science in Latin America and the exchange of ideas between scholars living in different countries are growing in exciting ways. I hope this book will contribute to these exchanges, and help advance our collective understanding of science and democracy around the world.

Acknowledgments

In the course of researching the material for this book and completing the manuscript, I have accumulated many debts of gratitude with colleagues, friends and family, and people in Chile who shared their time and experiences with me. This book began as a PhD dissertation at the University of California at Berkeley, and would never have come to fruition without my committee's guidance: David Winickoff, Kate O'Neill, Todd LaPorte, and Alastair Iles. Thank you for constantly challenging my assumptions while also trusting in me and the project, and for being willing to engage so deeply with the Chilean experience. I am also extremely thankful to Mark Healey, Alison Post, Paul Steinberg, Margaret Taylor, and Tiago Saraiva for their intellectual and professional advice over the years. My thanks also to many more Berkeley faculty who influenced me over the years, and to the students and postdocs affiliated with the Department of Environmental Science, Policy, and Management and what we affectionately called the "Christmas Center" (Center for Science, Technology, Medicine, and Society).

My colleagues at the University of California at Santa Barbara (UCSB) also provided intellectual and professional advice that helped bring this book to completion. At UCSB, I presented this work and received valuable feedback from participants in the Global Studies colloquium, Political Economy and Development Hub, Environmental Politics Working Group, Spatial Center, and summer writing group; I am grateful to everyone for their feedback at these events, including Gabriela Soto Laveaga, Alice O'Connor, Stephan Miescher, Simone Pulver, Sarah Anderson, Casey Walsh, Aashish Mehta, Jan Neverdeen-Pieterse, Kate McDonald, Vilna Bashi Treitler, Alison Brysk, Paul Amar, ann-elise lewallen, Werner Kuhn, Patrick McCray, and Barbara Walker. I want to recognize Peter Alagona in particular for taking the time to work through the book's main themes with me at a critical moment for this project as well as Eve Darian-Smith and Barbara Herr Harthorn for their career-defining advice. Finally, my thanks to Karen Lunsford, Jennifer Martin, Tammy Elwell, and Sylvia Cifuentes, who offered targeted feedback that strengthened this manuscript. Elizabeth Hofius assisted with the research that informs this book's conclusion.

I am also lucky to have a network of colleagues across the United States and South America who provided their time and thoughts to improving my work; thank you to Bhavna Shamasunder, Andy Gerhart, Chris Jones, Maite Salazar, Manuel Tironi, Eden Medina, Carlota McAlister, Sam Evans, Isabella Alcañiz, Laura Foster, Jessica Smith, Peter Benson, Mark Gardiner, Sebastian Ureta, Logan Williams, Gwen Ottinger, and Iván da Costa Marques. I presented draft chapters at many conferences and workshops, where I always received useful feedback; thank you to everyone who served as discussant or attentive audience member at the meetings of the Latin American Studies Association, 4S, American Anthropological Association, International Studies Association, and others. This book was also strengthened by my participation in the Contested Expertise workshop organized by Vivien Hamilton, Brinda Sarathy, and Janet Brodie in Claremont, California.

Beth Clevenger, Robert Gottlieb, and the staff at the MIT Press kept things running smoothly and recruited three anonymous reviewers whose careful reviews greatly improved this book. Audra Wolfe and Clark Miller provided sage advice for how to turn this into a readable manuscript; I am extremely grateful to them.

The National Science Foundation and a summer grant from the Social Science Research Council (guided by Ben Ross Schneider and Andrew Schrank) funded research for this book, and UCSB provided me with funds to focus on writing. I also used UCSB funds to hire Bingzheng Xu, with UCSB's Spatial Center, to draw the maps used in this book. These various sources of financial support made this book possible, and I am thankful.

I would like to thank everyone in Chile who participated in this project. In particular, I would like to thank a few individuals whose support helped shape the final manuscript: Antoine Lassagne, the Garrido and Schindele families, Michaela Heisig, Patricio Rodrigo, Ignacio Toro, and my housemates in Valdivia, Nuria and Loreto. Thank you for opening doors for me in Chile, sharing so much of your time, and caring deeply about the quality of Chile's environment.

Finally, I have some personal debts with family and friends. Special thanks to those friends who kept student life fun: Emilio Martínez-Velasco, Becky Tarlau, Freyja Knapp, Jelena Simjanovic, Diego Echecopar, Rosario Nava, Thuy Greer, and many more with whom I've shared across the Bay Area, Santiago, and Santa Barbara. My family lives distributed between Santiago, Oakland, and Buenos Aires, ensuring that there was never a dull moment. Thank you for keeping me company and helping me every time I asked for help. If the last three years of living with this project have been exceptionally happy, I owe that to my partner, Tristan Partridge. Both Tristan and my sister, María José, improved this manuscript by reading multiple drafts as well as engaging with me in conversation over many of the ideas and events discussed in the pages that follow. Thank you both so much. All remaining errors are my own, of course.

List of Acronyms

AOX	adsorbable organic halides
CEA	Centro de Estudios Ambientales
CEAZA	Center for Advanced Studies in Arid Zones
CENMA	National Center for the Environment
CIEP	Center for Ecology Research in Patagonia
CIPMA	Center for Environmental Research and Planning
Conaf	National Forest Corporation
Conama	National Commission for the Environment
Conicyt	National Science Agency
COPAS	Center for Oceanographic Research in the Eastern South Pacific
Corfo	Production Development Corporation
EIA	Environmental Impact Assessment
EIA Agency	Environmental Impact Assessment Agency
EPA	Environmental Protection Agency
EULA	Europe-Latin America Center
GAC	Gestión Ambiental Consultores
IFOP	Fisheries Research Institute
INFOR	Forestry Institute
ISA	infectious salmon anemia
JICA	National Aid Agency of Japan
NEPA	National Environmental Policy Act
NGO	nongovernmental organization
OECD	Organization for Economic Cooperation and Development
OLCA	Latin American Observatory for Environmental Conflicts
RAMA	Environmental Regulation for Aquaculture Act

SEIA	EIA Service
Sernapesca	National Fisheries Agency
STS	science and technology studies
Subpesca	Subsecretariat of Fisheries

Introduction: States and Knowledge of Nature

"This used to be a Japanese garden," Maria said as she guided me through the National Center for the Environment (Centro Nacional de Medio Ambiente, or CENMA). "That was the fountain, it was once full of water and even had carp in it. There were Japanese maple trees that changed colors with the seasons." I could hardly believe her. The soil we stood on was hard-packed and the grass like hay. The fountain was dry, with its concrete structure cracked and threatened by weeds. "In the garden you felt like you were in Japan, not here in Santiago," she continued, as we took in the dry, smoggy air of Chile's capital city. The National Aid Agency of Japan (JICA) funded CENMA in 1995 to develop scientific advice for environmental policy. The garden symbolized the two nations' shared commitment to environmental protection—a commitment that might have grown like carefully tended plants. But instead of a tidy Japanese garden, I stared out at a parched plot of dying plants. The garden had not been destroyed by one of Chile's notorious earthquakes. Rather, its distress resulted from the Chilean state's neglect of environmental science—a blind spot that continued to grow after Japanese support for CENMA ran out.[1]

The Chilean state's refusal to invest in scientific knowledge for policy making seems peculiar in the early twenty-first century. Scholars from a range of disciplines have observed that states have used scientific knowledge to pursue their colonizing, civilizing, military, and governance goals. European powers used science and technology to build their colonies, and in turn used those territories to advance their expertise (Adas 2014; Hodge 2007; Tilley 2011; Storey 1997). Since World War II, the United States has spent massive amounts of public funds on science, convinced that science is "the endless frontier" and crucial to promoting US economic and political interests at home and abroad (Krige 2008). Over the twentieth century in the United States and parts of Europe, the expansion of the state into welfare and planning brought experts—typically men trained in math, statistics, economics, engineering, and the natural sciences—into government, extending nineteenth-century positivism to the modern era (Hacking 1990; Porter 1995; Rueschemeyer and Skocpol 1996). Developing and newly sovereign countries also turned to science and technology, sometimes in nationalistic ways, to pursue economic growth as well as "civilize" populations considered poor and culturally backward (Cushman 2013; Prakash 1999; McCook 2002; Soto Laveaga 2009; Scott 1998; Hecht 2011). Governing elites felt it was especially important to know nature: nature is "too unwieldy in its 'raw' form for administrative manipulation," wrote James C. Scott (1998, 22) in *Seeing Like a State* (see also Hays 1959). He further cautioned, "If we imagine a state that has no reliable means of enumerating and locating its population, gauging its wealth, and mapping its land, resources and settlements, we are imagining a state whose interventions in that society are necessarily crude" (Scott 1998, 77).

Such enthusiasm for science reflects the belief that science has symbolic value in a liberal democracy (Brown 2009; Ezrahi 1990; Jasanoff 2005). In the abstract, science aims to deliver objective knowledge that transcends the material interests, cultural biases, or superstitious beliefs of individuals or identifiable groups. This was the intention of the Enlightenment and the later Progressive Era, when many had faith that experts could act as a "reality check" on the power of the oligarchy (Hays 1959). Science promises to remove personal discretion, subjectivity, or judgment from decision-making (Ezrahi 1990). Decisions would no longer appear arbitrary, but rather be the result of careful, reasoned debate based on scientific data. Contemporary industrialized liberal democracies have devised a variety of ways to organize science for policy; science retains its special authority on everything from the environment to public health risks, negotiated through institutionalized cultural practices that allow scientists to participate in decision-making, even as it has become clear that science cannot resolve all conflicts over policy or regulation (Bijker, Bal, and Hendriks 2009; Callon, Lascoumes, and Barthe 2012; Frickel and Davidson 2004; Guston 2000; Hilgartner 2000; Jasanoff 1990, 1995, 2004, 2005; Keller 2009; Lachmund 2013; Latour and Weibel 2005; Miller 1998; Miller and Edwards 2011; Owens 2015; Price 1965; Waterton and Wynne 2004; Yearley 1996). Though institutionalized models of scientific advice vary widely, they nonetheless cling to the old adage that *scientia potentia est—la ciencia es poder*: knowledge is power.

Knowing all this, why would Chile's new democratic leaders—at the helm of a natural resource–dependent economy and a society demanding environmental protections—imagine that the state could do without an environmental science center like CENMA? Even after receiving JICA's donation, the Chilean state let that capacity wither away. One possible explanation lies in the practical failures of scientific advice: science does not always provide complete, clear, or unequivocal policy answers on even seemingly technical issues, like calculating a sustainable fish catch or identifying the environmental and health impacts of specific pollutants (Jasanoff and Martello 2004; Sarewitz 2004). Another explanation might point to limited financial or technical resources or other constraints found in developing countries like Chile. In this view, Chile is too poor or "backward" to do science effectively—a view echoed in international studies of environmental regulation that assess how globalized policies work in different countries by comparing messy realities with idealized checklists of how they should work (e.g., Lee and Clive 2000). Yet another explanation points to a broader cultural shift toward skepticism of experts in response to the increasing complexity of real-world problems, proliferating sources of information and expertise, and declining trust in the capacity (and benevolence) of the state to maximize its citizens' welfare (Ezrahi 1990; Owens 2015). In Latin America, for instance, Marxist intellectuals have long regarded science with skepticism, believing it to be too concerned with bourgeois pursuits to be useful for the average citizen (Beigel 2013).

A fourth explanation, the rise of neoliberalism, is of special relevance for Chile, where neoliberalism had an early and powerful influence. Neoliberalism is a political ideology that privileges market-based solutions to collective needs over those that the state can provide. Its intellectual roots can be found in the work of economists Friedrich Hayek, Milton Friedman, and others (Plehwe 2009). Chile adopted neoliberal policies earlier and more thoroughly than elsewhere. Neoliberalism has guided Chilean governments since the mid-1970s, when General Augusto Pinochet—who ruled as a dictator from 1973 to 1990—turned for policy advice to economists at the Pontifical Catholic University of Chile (Fischer 2009; Valdés 1995). These economists became known as the "Chicago boys," a reference to their training and commitment to neoliberal policies developed at the University of Chicago. These policies include new property rights, privatization of state-run enterprises, smaller state agencies, and market-based regulations.[2] Beyond Chile, since the 1980s, states everywhere have been privatizing state-run entities and outsourcing services that state agencies used to provide in-house. This includes centers of knowledge production, like CENMA, that are disappearing in favor of think tanks as well as for-profit consultants and laboratories, with complicated consequences for governance (Carey 2010; Lave 2012; Mirowski 2011; Berman 2012).

Elements from all these hypotheses help explain the state's lack of vision for science on strategic issues like the environment in the years after Chile transitioned from dictatorship to democracy. Indeed, part of why Chile is such an interesting case for the study of scientific advice in a developing country is its reputation for expert-led governance (Medina 2011; Silva 2009; Ureta 2015). International organizations have repeatedly celebrated Chile's supposed reliance on experts as a leading factor in the state's shift from "developing" to "developed" country status (Stein et al. 2005). During the transition to democracy, individual scientists were outspoken about their hopes for science to have a greater voice in a future government, through new labs like CENMA and new environmental policies. Given all this, I expected Chile to be a "most likely" case to adopt modern—that is, science-based—environmental institutions and policies. And yet the opposite situation has occurred, suggesting that we as scholars need to rethink our understanding of the relationship between science and power in emerging neoliberal democracies.

This book examines the organization of science for policy in an emerging democracy through four semi-independent environmental conflicts that left a lasting imprint on Chilean politics and culture in the 2000s. Most theories of the relationship between science and power in modern democracies assume that science acts like a neutralizing agent in decision-making; it adds a dose of objectivity and realism that transcend social commitments. In contrast, I find that in Chile, many state officials see themselves as that neutralizing agent. Rather than science, they believe the state itself acts like a "neutral broker." As a neutral broker, the state does not claim to speak for the "common good." Instead, it brokers between private parties without necessarily seeking good policy outcomes and generally (but unsuccessfully) operates to minimize adversarial conflict. Scientists in Chile have participated in environmental decision-making in numerous ways, from consultancies to universities. Of these, CENMA's rise and fall points particularly well to the troubled character of science in contemporary Chilean political culture: what science means for regulation, how to define and organize it, and even when to appeal or defer to it are unsettled questions among Chilean actors. Moreover, the case of Chile—the epitome of a neoliberal state—shows the kind of state and society that are produced alongside a market for scientific advice. The neoliberal neutral broker state cannot advance official accounts that scientists or citizens find acceptable, and surrenders instead to tired, repetitive debates about authority, accountability, and trust. It thus runs into major difficulties when trying to balance economic growth and environmental protection.

Environmental activists and concerned citizens might find these patterns familiar. They have long warned that Chilean political and business elites' commitment to an export-led development model would have terrible environmental consequences (Altieri and Rojas 1999; Larraín 1999; Liverman and Vilas 2006; Quiroga and Larraín 1994). Even the Organization for Economic Cooperation and Development (OECD 2005) criticized Chile's environmental rules and regulations as too lax. By the late 2000s, two-thirds of Chileans lived in cities where air pollution regularly exceeds national air quality standards.[3] Coal and petroleum coke power plants dotted the north of Chile and many coastal areas. A nongovernmental organization (NGO) labeled five industrial ports "sacrifice zones" because of their toxic pollution—an appellation echoed even by the business press (Salas 2012). Entire rivers now run dry, like the Copiapó, which twenty years ago still flowed through the Atacama Desert. Activists blame the state for this environmental destruction, because dirty and resource-intensive industrial projects, like mines, dams, and industrial plants, must receive state approval to proceed. Environmental Impact Assessments (EIAs) require developers to submit a scientific review of the environmental effects of any industrial project prior to receiving construction and operating licenses. Scholars and activists frequently accuse the state agencies that grant EIAs of issuing permits without fully considering the scientific evidence regarding environmental impacts and their consequences for local communities (Carruthers and Rodriguez 2009; Sepúlveda and Villarroel 2012).

By 2011—during my fieldwork for this book—these and other grievances had driven Chilean citizens to the streets for demonstrations. In January, they protested changes to natural gas subsidies; in May, thousands in cities across the country marched against the hydroelectric project HidroAysén,

analyzed in this book; in June, students organizing on behalf of public education paralyzed national politics. The protests expressed a general malaise that
had been accumulating for some time: that the state faced a crisis of legitimacy
because it lacked the capacity to answer citizens' demands (Figueroa 2013;
Garcés 2012; Ureta 2015). These included demands for a cleaner environment along with a more equitable and sustainable development model—one
that did not consistently favor the interests of industry. For many citizens,
the state's commitment to industrial interests over the environment was
ratified in every EIA decision, because EIAs require state agencies to balance
environmental values against economic benefits. EIAs thus became prime
objects of conflict, pitting not only economic growth against environmental
values but also a procedural view of democracy against demands for a more
capable state that could respond to citizens' demands.

To synthesize these struggles over the status and value of science for
policy, I rely on two anchoring devices: the state as "empire" and its opposite, the state as "umpire." James C. Scott (1998) has written at length about
the state as empire, referring to a state that is confident in its scientific and
technical capacities, and hence can "make the world it wants to govern." By
contrast, the umpire state sees itself as a broker between competing parties
that produce their own knowledge claims. The ideal of the umpire state is
rooted in the belief that markets aggregate and process information better
than governments, which cannot possibly know everything necessary to
manage the economy (Hayek 1945). Friedman later applied this argument
to criticize expert advice on everything from railroads to public housing, and
advocated for eliminating guilds, bars, and other methods used to certify
professional competence. In *Capitalism and Freedom*, Friedman (1962, 25)
used the term "umpire" to advocate for a state that would "provide a means
whereby we can modify the rules, to mediate differences among us on the
meaning of the rules, and to enforce compliance with the rules on the part
of those few who would otherwise not play the game."

Friedman (1982) also famously described Chile as an economic and
political "miracle" because it had aggressively pursued his vision of the
state. Despite their close links to Chilean policy makers and personal
involvement in Pinochet's regime (Fischer 2009), neither Hayek nor Friedman directly shaped Chile's environmental or science policies. Nor am I
suggesting that Chile or any other state has definitively shifted between
two static modes—that of an empire or umpire. Rather, these are alternative

ideological philosophies that have, in uneven yet powerful ways, guided officials and politicians in their relationships with scientists and environmental decision-making. Given how far Chile has gone in implementing neoliberal principles, its experiences with EIAs and environmental politics can offer valuable clues as to how Friedman's notion of the umpire state might work in practice, or at least, what happens when these neutral broker ideals meet alternative visions of the state that hold reasoned judgment, defense of the common good, pluralism, local control, and independence from business interests in high esteem.

Methods and Cases

This book advances scholars' understanding of science and power in a democracy by looking at how science was organized in a country on the Latin American "periphery" as it transitioned from dictatorship to democracy in a neoliberal context. I tell this story through the experiences of emblematic labs like CENMA, and by following science through the four environmental conflicts that have most raised public anger and rattled Chilean citizens' confidence in the state since democracy was restored (Bustos, Prieto, and Barton 2015). These are:

1. Salmon farming: The rise of this industry transformed small fishing communities into the world's second-largest suppliers of salmon, until an epidemic in 2007–2008 killed thousands of fish and led the industry to lay off twenty-five thousand workers.
2. Celco Arauco's Valdivia paper and pulp mill: Soon after opening, local residents accused this mill of polluting the surrounding wetland, forcing thousands of black-necked swans to flee the area or die from starvation.
3. Barrick Gold's Pascua Lama mine: The Chilean state approved a plan by Barrick to move three small glaciers to "protect them," but then, following public outcry, backtracked. As of 2017, construction remains paralyzed after a string of environmental violations.
4. Endesa's HidroAysén hydroelectric dam project: Intended as a flagship project for the state, opposition to HidroAysén rallied environmentalist sentiment across Chile like never before, as many opposed the idea of industrializing Patagonia—a region famous for its unspoiled landscapes—in exchange for industry profits that would benefit only a few.

Figure 0.1
Map of Chile, with the capital city (Santiago) and the cities
closest to each conflict analyzed in this book.

The events analyzed in this book span four industrial sectors (salmon farming, forestry, mining, and energy), three government administrations (presidents Ricardo Lagos, Michelle Bachelet, and Sebastián Piñera), and four regions (Los Lagos, Los Ríos, Huasco, and Aysén). EIAs are the thread that ties these conflicts together. Developed in the United States, EIAs have by some estimates become the most widely emulated US policy in the regulatory playbook and preferred policy for managing natural resource use by connecting scientific evidence with public input in a highly regulated bureaucratic procedure (Greenberg 2012). International comparative studies of EIAs abound. Typically, these compare how variables such as analytic capacity or bureaucratic autonomy influence compliance, stability, and other desired outcomes across countries (Espinoza and Alzina 2001; Glasson, Therivel, and Chadwick 2012; El-Fadl and El-Fadel 2004; Kolhoff, Driessen, and Runhaar 2013; Marara et al. 2011). Comparative studies like these treat EIAs as a model: a "best practice" approach to environmental management that can be replicated worldwide. JICA (2003) followed a similar logic in funding CENMA and labs like it in China, Thailand, Mexico, and Vietnam.

Despite their popularity, EIAs are hardly the only model available for governments to emulate. STS scholars have found that liberal democracies use a variety of ways to organize scientific advice; there is no one tried-and-true model. This variation reflects political culture, understood as "the systematic means by which a political community makes binding collective choices" (Jasanoff 2005, 21).[4] In *Designs on Nature*, Sheila Jasanoff further argues that when it comes to producing and applying knowledge in collective choices, societies have different preferred methods for producing and validating knowledge that reflect existing institutions as well as relatively stable and shared normative commitments. She calls these *civic epistemologies*. As a result, we as scholars should expect that how EIAs work in practice will vary depending on the broader civic epistemology.

In what follows, I analyze four conflicts as different chapters in a single ongoing sociotechnical controversy over how best to organize scientific advice for environmental management in an emerging, neoliberal democracy. My analytic approach draws on Jasanoff's civic epistemologies framework to focus on boundary work and modes of institutional reasoning to capture the regularities in how the state tries to evaluate and legitimate large industrial projects, and how different actors challenge these efforts. My objective is to understand "the self-perpetuating normative commitments that give societies a claim to coherence and solidarity even in the face of shocks and change" (Jasanoff 2005, 40). To achieve this goal, my analysis focuses on actors' efforts to draw boundaries around science in order to defend (or attack) its symbolic value in society (Gieryn 1999), and the ways in which institutions "think" and choose to express those "thoughts" to citizens.[5] In a book on contemporary environmental conflicts in Chile, the institutions that matter—and that I studied—are state agencies, public and private research centers, corporations, and environmental NGOs that participate in as well as contest EIAs. Of interest too are the actors who move to make and defend the boundaries of science, particularly the scientists, consultants, state officials, and company officials who played pivotal roles in each of these four conflicts along with leaders from the activist and political worlds. In all, I conducted about a hundred semistructured interviews in Chile between November 2010 and July 2011 (and during shorter visits in 2014 and 2015) in the towns where these conflicts took place: La Serena, Vallenar, Santiago, Valdivia, Concepción, Puerto Montt, Puerto Varas, Coyhaique, and Cochrane.

I selected these four cases for what they have in common: they all involve natural resource–based extractive industries in rural areas and were widely perceived to have put democratic era institutions to the test. My first encounter with each conflict was not as a researcher but rather as a Chilean citizen who, though living abroad since childhood, has had the privilege to visit at regular intervals. As 2010 started, no conflict so moved society as did these four, with images of dying salmon and swans as well as shrinking glaciers flashing regularly across TV screens. Plenty of other environmental controversies existed. Scientists I interviewed often suggested I look instead at conflicts like the Alumysa aluminum plant, Trillium forestry project, or effects of mining on Lake Chungará, because scientists opposed these; activists referenced emblematic conflicts like the Ralco and Pangue hydroelectric dams. The regulations that applied to those conflicts still reflected dictatorship era constraints. By contrast, nearly everyone I interviewed agreed that the four selected conflicts were the first to challenge the environmental framework law passed after the return of democracy.

I also selected these conflicts because the political and economic power of the groups involved in them varies. In some, up-and-coming scientific communities, like glaciologists or marine biologists, played important roles alongside groups with long-standing political influence like engineers and, to a lesser degree, biologists. Different kinds of companies participated. Salmon farmers at the time of the epidemic had a strong industry association and espoused an antigovernment rhetoric that set them apart, politically and socially, from Chile's traditional elite (Bustos 2015). The opposite is true of Celco Arauco, one of Chile's largest and best politically connected family-owned conglomerates. Barrick Gold of the Pascua Lama mine is a Canadian multinational that compared to other mining companies, had weak ties to government due to its recent arrival. And Endesa, the company behind HidroAysén, is a well-connected, formerly state-owned enterprise, now owned by Italian and Spanish capital. This variation is important because in Chile, multinational and family-owned companies wield a lot of power (Khanna and Yafeh 2007) but through different channels (Ross Schneider 2004). The variation in ownership allows me to attribute observed patterns to routines and cultural mores, rather than to the influence of an especially powerful company or set of companies, like multinationals or family-owned businesses, or the incompetence or corruption of a specific official or state agency. Inversely, looking at more conflicts would have required quantitative, not interpretive, methods.

Contests over the status and value of science are not only intellectual exercises. Instead, as in Chile, they can shape what kinds of capacities and responsibilities state agencies develop, because—following the coproductionist idiom—the ways in which a society choses to know about the world both make and are made by the ways in which we collectively chose to live in the world (Jasanoff 2004). The analysis that follows shows ongoing struggles to define science for regulatory purposes, identify who practices it, and understand when to use it as a political resource. Chilean state agencies had trouble demonstrating to citizens the "technical" or "scientific" foundations of their decisions, in part because state officials were equivocal about what technical should mean in environmental decision-making. For some, technical meant scientific, while for others it meant "according to the rules." Chilean scientists likewise struggled to defend their professional credibility from accusations that they held conflicts of interest or did poor quality work. Despite making an effort to do so, scientists struggled to act as a critical community that has regular access to institutions and policy makers. The unsettled status of science is also reflected in the analytic capacities state agencies developed, which in turn elicited skepticism among such different users of environmental data as peers in state agencies, environmental watchdogs, and regulated industries. Thus, even when some agreement that science should play a role in regulation existed, neither universities, consultants, nor the state were sufficiently authoritative to impose their particular model of science on environmental politics.

That Chile presents a puzzling case of a modern state that appears unaware of the symbolic value of science in environmental governance holds true across these conflicts and others that have occurred since the transition, as is also evident in the trajectory of CENMA along with other public and private labs. When I selected these cases, I did not anticipate that in each I would find a different publicly funded research lab whose advice would be pushed out through different forms of boundary work and institutional reasoning. Neither did I expect each conflict to become an environmental victory of sorts. Since 2011, new regulations have been applied to salmon farming; a Valdivia court found Celco Arauco guilty of polluting the wetland; the authorities forced Barrick to protect the glaciers adjacent to Pascua Lama; and the government canceled HidroAysén. My analysis builds on the growing literature on how Chile's development policies, before and since the adoption of neoliberalism, have produced significant inequalities and environmental harm (Bustos, Prieto, and Barton 2015; Gerhart 2017; Klubock 2014; Tinsman 2014; Winn 2004). But by casting my gaze across industry sectors—rather than focusing on one sector or conflict, as is more common—this analysis underscores the extent to which these were not just conflicts over natural resources but rather moments in a larger controversy over how decisions about resources should be made and by whom (Hays 1959).

This was further driven home by the surprises that made 2010–2011 a pivotal year for environmental politics. In January, Piñera became the first right-wing president elected since 1958, and his government had to implement a major reform to the environmental framework law: the National Commission for the Environment (Comisión Nacional de Medio Ambiente, or Conama)—a small environmental coordinating agency created during the transition to democracy—was replaced with the Ministry of the Environment, an enforcement agency, and the autonomous Environmental Impact Assessment Agency (EIA Agency). Although education and environmental quality had dogged every president before him, public discontent with the Piñera administration soon reached extremely high levels. One year into Piñera's presidency, opposition to HidroAysén drew the largest demonstrations seen in Santiago since 1988, when people mobilized to end the dictatorship. Weekly anti-HidroAysén protests were then eclipsed by student demonstrations, which in June attracted as many as a hundred thousand people. My interviews took place in this context of reform and protest, which kindled reflection and openness about the shortcomings of "the old Conama" as well as hopes for the new institutions. One pattern I soon observed was that despite the apparent victories and reforms, interviewees expressed such low confidence in the state's capacity, or inversely, such high distrust of the state's intentions and competence, that few gave state agencies any credit for their efforts to undertake careful EIAs. This book explains why this is the case.

The Symbolic Value of Rules

In 1980, the Pinochet regime passed a new constitution, still in place today, that draws significantly from Hayek's 1960 book, *The Constitution of Liberty*. It protects a neoliberal conception of freedom as "intrinsically connected to private property, free enterprise, and individual rights" (Fischer 2009, 327). Although it guarantees civil liberties and the "right to live in an environment free of pollution," in practice the rights to free enterprise and private property have priority (article 19). The 1980 constitution thus limited citizenship rights and created a strong presidential office supported by a powerful group of ministers, while limiting the discretionary authority of state agencies (Garcés 2012; Wormald and Brieba 2012). An analyst at Chile's Constitutional Tribunal writes that article 19 guarantees what he calls the "subsidiary principle": state action is restricted to that which is expressly allowed by law, whereas private parties can act in any way not expressly forbidden to them (López Magnaso, 2012, 14). Chilean scholars as well as the state officials I interviewed often cite this principle to explain how state officials make decisions. This principle expands the private sector and restricts the public, limiting state agencies' responsibilities to enforcing "the rules of the game."

In reviewing a new industrial project's environmental impact study, several state officials explained to me that their role was to "draw the lines on the soccer pitch" or "be the net on the tennis court." State officials described themselves as umpires, referees, and even inanimate objects like lines and nets that communicate the rules of the game. With these terms, they justified their decisions by appealing not to the symbolic value of science but instead the symbolic value of rules: in a state where the subsidiary principle guides officials, the promise of objective, impartial, depersonalized governance comes through an idealized, abstract notion of the rules rather than from science. The imaginary of the state is not one where science and its promise to deliver an objective "reality check" reigns supreme, or even one where experts play the role of "honest brokers" (Pielke 2007). Instead, in the neoliberal state guided by this subsidiary principle, many officials express a commitment to a normative imaginary of the state itself as a neutral broker in which rules have symbolic value.[6] State officials who subscribe to this view therefore do not feel responsible for advancing an agenda like environmental protection. In this kind of state, no one speaks for the common good.

Chilean actors strategically mobilized these polar ideologies—a vision of the state as an empire with scientific capacity or a neutral umpire that enforces the rules—to protect themselves from the outcomes of their decisions or advance their goals. The chapters that follow draw out the practices and consequences of the neutral broker imaginary of the state for how science is organized for environmental policy and regulation. When state officials choose to act as neutral brokers, this has consequences for how the environment becomes known and represented in EIAs—chiefly, as a fragmented and uncertain space. Some officials appealed to the symbolic value of rules to reduce their discretion, and thus avoid having to weigh environmental values against competing economic goals, or having to make careful judgments based on the needs of specific communities, ecosystems, or projects. Another consequence of this vision of the state is that agencies developed limited in-house capacities; instead, officials had to rely on external scientific advisers and off-the-shelf scientific advice, and were rewarded not for producing official accounts of nature but rather for brokering fragile agreements between industry, citizens, and the state regarding natural resource use. But many actors working both inside and outside state agencies contested this imaginary. Some retorted that in principle and practice, the state cannot be neutral because it is responsible for environmental stewardship and protecting communities' quality of life. Others criticized that neutral broker state agencies would forever lack sufficient analytic capacity to produce the authoritative, nonnegotiable representations of nature that are necessary for good environmental governance.

Symbolic power aside, in practice neither science nor rules deliver the objectivity or neutrality they promise. And whether they lean toward the empire or umpire model, no state has consistently delivered positive environmental outcomes. Since the 1970s, however, states worldwide have been under increasing pressure to act more like umpires than empires. Not only have property rights and market-based mechanisms been expanded, but in reordering the funding and organization of science, neoliberalism is changing how states produce, validate, and apply knowledge claims in decision-making along with the relationship between "experts" who claim to speak for nature, and the publics and entities they seek to represent. Critiques of neoliberal governance have demonstrated its poor results for the environment and social justice. In attending to the forms of knowledge making that neoliberalism promotes and analyzing their consequences for environmental governance, this book captures a vital part of the story that those critiques often miss. This includes an understanding of the tensions that occur over how best to organize knowledge production and state capacities in a neoliberal state, and the kind of social order, with its inclusions and exclusions, that is coproduced alongside this state.

The Book's Structure

The first two chapters set the stage for the four conflicts, analyzed in chapters 3–6, which are organized from the broadest level of scientific advice to the most specific: that is, from ecosystem monitoring down to the production of a specific study for a particular project. Chapter 1 describes the new regulatory institutions set up during the transition to democracy, and establishes the importance, functioning, and disputes engendered by the rules and regulations governing EIAs in Chile. Chapter 2 shifts the focus to Chilean science: it examines how Chilean universities fared through dictatorship and the transition, and then juxtaposes two new environmental science laboratories—one of which is CENMA—that were supposed to train scientists and administrators to work in and for the new democracy's institutions. This chapter charts the rise of a privatized model of scientific advice at the expense of a publicly funded one.

Chapter 3 examines the state's capacity to analyze the water quality at salmon farms—a seemingly necessary step for anticipating environmental and health risks to the salmon as well as other species resulting from farming practices. Aquaculture-specific EIA regulations require all farms to regularly monitor ocean conditions, which is work done by for-profit consultants. This case demonstrates how market-based science can breed distrust and also shows the shortcomings of the (neutral broker) state's strategies for controlling distrust—apply more and more precise rules. This situation reached a crisis in 2007–2008, when an epidemic killed millions of fish and led salmon companies to fire twenty-five thousand workers. In response, the state changed its approach in a way that departs from the neutral broker state ideal: rather than try to improve public trust in environmental information through more precise rules, it regulated the market for consultants.

Chapter 4 examines how environmental science fares when the state needs to enforce the permits and conditions it established during the review of an EIA. The specific case, which involves litigation against a paper and pulp mill, was the first test of Conama's capacities to improve a project through its EIA. Despite Conama's efforts, soon after the mill began operating, nearby residents accused it of polluting the river and causing the environmental collapse of the wetland downstream from the plant. The results were devastating: thousands of black-necked swans fled the wetland in search of food elsewhere, and many died. Assisted by rival scientific teams, the state and the company fought an eight-year court battle in which what counts as science good enough to prove environmental harm was highly contested.

Chapter 5 looks at efforts by civil society to use science to challenge EIA studies during Conama's assessment of the project. The company Barrick Gold wanted to build a large gold mine, called Pascua Lama, at five thousand meters above sea level, where its activities threatened three glaciers. Alarmed, the agricultural community in the valley below mobilized against the project. It called glaciologists to help it transform the glaciers into a matter of national concern. This chapter explores whom scientists speak for in EIAs, how productively they can challenge claims made by other actors, and what influence their participation has on the state's efforts to broker an agreement between interested parties. This case showcases deep ethical divisions between interest groups, with consequences for who participated in the controversy, and how claims came to be seen as certain or uncertain.

Chapter 6 examines the production of ecological science for Hidro-Aysén's EIA. With five dams on two rivers in Patagonia, HidroAysén was supposed to be a flagship national project that would bring development to this remote area and deliver cheap electricity to consumers in Chile's cities and mining industry. The chapter illustrates how unevenly science participates in the making and assessment of an EIA. Administrators and politicians blurred the boundaries between scientific, technical, and political criteria to promote their rival visions of the state as empire and umpire. Three years after the EIA Agency first approved HidroAysén, the government withdrew the project's EIA permit, arguing it was a "bad" project and raising the question, Are there projects a neutral broker state cannot legitimate because of their size and complexity?

The book's conclusion reflects on the kinds of state capacities the umpire or neutral broker state favors, and the consequences of this for Chilean environmental politics in light of ongoing efforts to improve how EIAs operate. The book finishes with a reflection on neoliberalism and its relationship to science in the world of the twenty-first century. As I was finishing this manuscript, UK citizens voted to leave the European Union and US citizens elected Donald J. Trump as president. Both elections unleashed commentary on the rise of "post-truth" democracies, influence of "alternative facts," and demise of liberal ideals in which science has high symbolic value. The environment again became a key site of conflict—so much so that scientists in the United States marched on Washington, DC, with placards reading "facts are our friends." Though not directly related, the events that happened in Chile hold cautionary lessons for societies (and environments) where the status and value of science is in dispute.

1 Environmental Protection, the Chilean Way

In 1990, President Patricio Aylwin assumed office and restored Chile's liberal democracy after seventeen years of military dictatorship (table 1.1). Fearful of causing economic or political instability that might jeopardize the new democracy, Aylwin's government acquiesced to several policies from the dictatorship era. The 1980 constitution remained in place, and the Senate stayed in the hands of promilitary figures.[1] The democratic governments that governed Chile between 1990 and 2010—led by the Concertación parties, a center-right coalition—embraced a style of neoliberalism similar to that being promoted by the World Bank and International Monetary Fund at the time. Economic growth was strong and politics stable. Between 1990 and 1996, the economy grew over 7 percent per year, with poverty falling from 39 to 23 percent (Winn 2004). Chile gained accolades from the international development community (Stein et al. 2005; Wormald and Brieba 2012), eventually gaining membership into the OECD in 2010. On that occasion, the OECD reported, "Chile's acceptance for OECD membership marks international recognition of nearly two decades of democratic reform and sound economic policies. For the OECD, Chile's membership is a major milestone in its mission to build a stronger, cleaner and fairer global economy. The *'Chilean way'* and its expertise will enrich the OECD on key policy issues" (Chile Signs Up 2010).

But not all has been well in the Chilean republic. Throughout the 1990s, activists and scholars denounced the social, economic, and environmental consequences of Chile's neoliberal model. Neoliberal laws creating property rights over water, fisheries, and forests, for example, fostered new industries that generated company profits at the expense of communities forced into low-wage jobs in increasingly polluted environments (Winn 2004). By the

Table 1.1
Timeline

1973	Military coup led by Augusto Pinochet
1980	"Constitution of Liberty" approved
1990	Democracy restored under President Patricio Aylwin
1994	Environmental framework law approved, establishing Conama and EIAs
1995	CENMA is created
2000	Pascua Lama gold mine: EIA assessment begins
2004	Celco Arauco's Valdivia mill: crisis strikes
2007	Salmon farming: crisis strikes
2008	HidroAysén: EIA assessment begins
2010	Chile joins OECD. Environmental framework law is reformed. Conama closes in favor of the EIA Agency, Ministry of the Environment, enforcement agency, and environmental tribunals

mid-2000s, despite declines in absolute poverty, economic inequality had grown dramatically from already-high levels: whereas in 1990 the top 5 percent of the richest households earned 130 times more than the poorest 5 percent, in 2003 that number had increased to 209 times more (Figueroa 2013). Though they may not be as poor as they used to be, working families are today saddled with debt from payments for education, health, and pensions.

Many Chilean scholars blame the 1980 constitution and timidity of the transition from the dictatorship for these inequities. Democracy was restored, but the rights of citizens along with the scope and role of the state remain limited (Garretón 2003; Moulian 2002). In contrast to the OECD's triumphant depiction of the Chilean way, historian Mario Garcés (2012, 23; emphasis added) called the transition "an agreement done '*a la chilena*,' a pact amongst elites, without the people, to reinstate the old 'procedural state' and a restricted democracy." These tensions between economic growth and inequality, and political stability and accountability, play out in the increasing number and visibility of environmental conflicts (Bustos, Prieto, and Barton 2015).

In his classic work *Contemporary Chile: Anatomy of a Myth*, Tomás Moulian (2002) argued that a gap exists between official discourse—the myth of Chile's success—and Chileans' lived experiences with inequality and injustices. This gap is captured in the triumphalist "Chilean way," expressed in English even in Chile, and its popular doppelgänger, "a la chilena." While

Chileans like to joke about doing things a la chilena—that is, dishonestly, cheaply, and without concern for others—elites use the phrase to assert the nation's modernity.[2] President Ricardo Lagos (2000–2006) used the Chilean way to describe what he considers a distinctly Chilean form of social democracy (Lagos, Hounshell, and Dickinson 2012, chapter 8). And after rescuing thirty-three trapped miners in a technologically sophisticated operation that captured the world's attention, President Piñera (2010–2014) similarly touted "the Chilean way of doing things, which means that it is a country where everything really works" (McElroy 2010; Benítez 2010).[3] Between the English-language Chilean way and vernacular a la chilena lies a tension in political culture between how the state *should* operate versus how it *does* operate. Environmental conflicts forcefully express this tension as they reveal different groups' efforts to strengthen the state's capacity—to either protect the environment, support a strong economy, or do both simultaneously—within and against the dominant political and normative commitments as well as material realities of the Chilean polity.

At issue are competing visions of the state: Is it strong and capable, as the triumphalist Chilean way suggests, or do state agencies operate a la chilena—that is, not as the serious, professional entities that national and international elites tout them to be? As a process that requires state officials to carefully balance environmental values against competing goals, EIAs provide a crucial site for investigating the tensions between the ideals of the Chilean state and how it works in practice. This chapter introduces Chilean environmental institutions after the transition to democracy—a moment when EIAs rose in importance—and then discusses EIAs in light of the scholarly literature on capacity building and scientific autonomy in Latin America. A "good" EIA decision requires a state with the scientific and intellectual capacity to assess the diversity of information the process generates. Despite global support, capacity building remains difficult to do, as the events in this book show. Arguments for capacity building in Chile have neither displaced neoliberal values nor addressed skepticism of foreign expertise rooted in the trajectory of Latin American science. The question thus remains, How to build real, effective, and accountable local capacities? This chapter argues the answer lies in cultivating "critical communities" that include scientists into routines and procedures that allow them to participate in the vetting, probing, and making of state policies as well as decisions.

EIAs and the Rise of Modern Environmental Management

By 1994, at the end of President Aylwin's term as Chile's first democratic president in nearly two decades, Congress had approved an environmental framework law—number 19.300—that established Chile's first modern institutions dedicated to environmental management. The framework law represented radical change from the environmental policies under the dictatorship: it recognized the need for environmental protections and created institutions whose job it would be to work toward that goal. But the environmental framework law also reflected neoliberal principles dominant at the time, including a preference for small state bureaucracies, limited regulation, and reliance on market mechanisms to guide state and private sector behaviors (Camus and Hajek 1998; Carruthers 2001; Nef 1995; Silva 1996). These commitments also influenced sectoral policies regarding water (Bauer 1998), forests (Klubock 2014), agriculture (Tinsman 2014), aquaculture (Barton and Floysand 2010), and electricity (Tironi and Barandiarán 2014).

As the name implies, the 1994 environmental framework law created the basic institutions and processes for environmental management in Chile. A small coordinating agency, called Conama, would ensure that ministries and state agencies considered the environment in their decisions and interventions. The framework also required that every industrial project undergo an environmental impact assessment, managed by Conama, before it could begin. Scholars have contended that Chile implemented EIAs—a global policy in use in all but a handful of countries (Pope et al. 2013)—in particularly neoliberal ways (Tecklin, Bauer, and Prieto 2011). By the late 1990s, EIAs had absorbed Conama officials' time and attention. The agency received EIAs from private sector actors seeking approval for their activities, distributed the studies to sectoral agencies for evaluation, held required public participation sessions, and collated the results into a report for the final evaluation committee to consider. Outside of managing and granting EIAs, Conama was a small, underfunded agency with few legal responsibilities (Silva 1994, 1996). For instance, Conama could not carry out unannounced inspections nor could it issue recommendations for environmental standards without first undertaking a years-long public review process and obtaining the blessing of its top advisory body.

By the 2000s, EIAs were so central to Chilean environmental politics that the word turned into a noun—Chileans referred to the "*seización*" of environmental politics (the *s* stands for "service," as in the EIA Service (SEIA), the administrative unit within Conama that managed EIAs). When a new Environmental Enforcement Agency began operating in 2013, its main task was to police the more than eleven thousand EIA permits that Conama had issued since 1994. Beyond that, the enforcement agency had just thirty pollution control measures and nine decontamination plans to enforce.[4] In short, the stakes of EIAs are high because Chile has few other laws, like pollution standards or land use policies, for regulating the environment (OECD 2005).

EIAs also became sites of conflict because they confront different visions of how a democratic state balances competing interests. Their shortcomings aside—and there are many—EIAs provide a particularly transparent window into how the state makes decisions. This is best illustrated by retracing the origins of EIAs in the 1960s' environmental movement in the United States, from where EIAs then spread globally. The United States' National Environmental Policy Act (NEPA), signed into law in 1970, was the first such law to require EIAs. The objective was to "legislate environmental awareness" by forcing state officials working at every office involved in the planning and licensing of infrastructure projects to consider, before approval or construction, a new project's environmental impacts and how these might be lessened. NEPA's promoters saw EIAs as revolutionary because they democratized expertise: before NEPA, these decisions were made by officials relying on specialized expertise drawn from within their agency, meaning that they were effectively shielded from knowledge claims brought by outsiders (Caldwell 1982; Greenberg 2012; Milazzo 2006). With NEPA, state officials could no longer justify a project solely on their agency's internal knowledge, but instead had to show that they had considered information provided by a host of groups, including affected citizens, Native American tribes, NGOs, scientists working at universities, consultancies, think tanks, other agencies, and others. NEPA thus simultaneously democratized expertise and consolidated its place in the decision-making process. And it channeled this array of participants into routine procedures.[5]

NEPA was perhaps particularly well suited to the United States, where a plurality of mobilized NGOs, civil society groups, and scientifically trained individuals working in many different kinds of organizations already existed.[6] Nevertheless, by the 2000s, NEPA-style regulation had become a global standard (Hironaka 2002), adopted by countries far and wide, regardless of the level of political and social mobilization around environmental or scientific issues. Endorsed first by the European Union in 1985, then the World Bank in 1989, and finally by the 1992 Rio Earth Conference, governments worldwide adopted EIAs to promote sustainable development (Cashmore, Bond, and Cobb 2007; Pope et al. 2013; Lee and Clive 2000). In its operational directive requiring them, the World Bank (1989, article 2) supported EIAs because they promised to ensure that "development options are environmentally sound and sustainable and that any environmental consequences are recognized early in the project cycle." In the early 1990s, governments wanted sustainable development policies, premised on the belief that "environment and development were complementary rather than contradictory categories" (Najam 2005, 249). EIAs provided a managerial approach to reconcile growth and environment—one of the greatest challenges developing countries face (ibid.). Within just a few years, the World Bank began to require borrower countries to incorporate EIAs into their policy making.

Despite the popularity of EIAs, it is difficult to assess their quality or effectiveness. Early studies of EIAs were interested in understanding the global diffusion of the policy, including identifying the conditions most likely to yield effective EIAs that adequately account for environmental impacts. These studies generally found that EIA practices fell short of the theoretical or procedural expectations that had been vested in them. The data used in EIAs were frequently incomplete or incorrect, human resources were often lacking, and the sustainability goals supposedly underlying the process were vague or parochial (Owens and Cowell 2002; De la Maza 2001; Clausen, Vu, and Pedrono 2010; Ortolano, Jenkins and Abracosa 1987; Nardini, Blanco, and Senior 1997; Bojórquez-Tapia and García 1998; El-Fadl and El-Fadel 2004; Glasson and Neves Salvador 2000). Corrective measures ranged from strengthening environmental quality standards and land use laws to diversifying who participates in EIAs, including calls for more involvement from local scientists.

Frustrated with the practical shortcomings of EIAs, scholars turned their attention to the potential of EIAs to transform society in more substantive ways, such as by opening spaces for dialogue among diverse publics and raising the profile of environmental considerations among state agencies (Holder 2004; Owens, Rayner, and Bina 2004; Cashmore, Bond, and Cobb 2007; Pope et al. 2013). Early studies of socioenvironmental conflicts around the environmental approval process argued that EIAs should integrate demands for environmental and procedural justice to challenge dominant power relations in government (Cashmore and Richardson 2013; Rozema et al. 2012). More recent work has highlighted the need for EIAs to consider alternative future scenarios—a prospect made more complex because of climate change, acute environmental pressures, and increased global interdependencies (Retief et al. 2016).

Missing from these studies is an analysis of the state, including the standards and logics by which a state operates and is held accountable—in other words, an analysis of the kinds of democratic states produced alongside EIAs. During NEPA's early days in the United States, the courts defined "good" EIA decisions in a way that put the state front and center. In the first verdict interpreting NEPA, in 1971, Judge J. Skelly Wright wrote that a good EIA decision results from a "fine-tuned and systematic balancing analysis" that weighs environmental values against other goals. He warned state officials that to achieve such an analysis, agencies could "not simply sit back, like an umpire, and resolve adversary contentions at the hearing stage. Rather, [the state] must itself take the initiative of considering environmental values at every distinctive and comprehensive state of the process beyond the staff's evaluation and recommendation" (cited in Milazzo 2006, 137).[7] In short, procedural specifics aside, a good EIA decision required a proactive, reflexive state, with officials capable of considering the information supplied to them with care, concern, and ability.[8] A passive umpire state, in which officials aspire to be neutral brokers, simply did not meet the judge's standard. Chile has, unfortunately, frequently been such an umpire.

Building State Capacities

Scholars have written volumes recommending technical fixes to EIAs, from changes in procedures, to greater legal clarity and more knowledgeable state officials, to adaptations for local realities (Cashmore, Bond, and Cobb 2007; Greenberg 2012; Pope et al. 2013). Most of these studies focus on the issue of "capacity." Early on, the World Bank (1992, 68) concluded that "lack of institutional capacity [was] the most serious impediment to improved environmental management" through EIAs. In response, the bank issued loans for EIA training and capacity building, including US$11.5 million given to Chile in 1993.

To the World Bank (1992, 68), limited capacity meant "inadequate or unenforced laws for environmental protection, weak or nonexistent environmental institutions, poorly trained staff, incomplete data collection and analysis, and poor monitoring and enforcement of existing laws and regulations." Scholars of EIAs, for their part, have emphasized the need to strengthen public sector human resources through training and good employment conditions, including political autonomy for state officials (Ahmad and Wood 2002; Doberstein 2004; Farrell, VanDeveer, and Jager 2001; Kolhoff, Driessen, and Runhaar 2013; Lee and Clive 2000; Marara et al. 2011; Pope et al. 2013). In practice, capacity-building programs have concentrated on funding research centers (like CENMA) along with formal training and monitoring activities (Goldman 2005; Grindle 1997; VanDeveer and Dabelko 2001). In 2005, for example, the OECD recommended that Chile improve its environmental performance by, among other things, strengthening environmental institutions, regulatory frameworks, and compliance and enforcement capacity; developing national indicators to measure environmental performance; expanding its knowledge base of existing environmental conditions; and strengthening state agencies' impact assessment and enforcement capacities in mining, forestry, and aquaculture.

Although the concept of capacity building receives widespread support, it has been difficult to implement (Grindle 1997; Miller 1998; VanDeveer and Dabelko 2001). A case in point is Chile's transition era environmental institutions. Like most capacity-building efforts studied by scholars, it involved international aid: JICA supported the creation of CENMA, and the World Bank helped fund Conama. Despite their achievements, neither CENMA nor Conama met their international funders' goals (Ruthenberg 2001; JICA 2002) or Chilean citizens' expectations (Silva 1997). The reasons have to do with the dominance of neoliberal values, resistance to foreign intervention, and certain limitations inherent to these kinds of policy efforts; these are all relevant to other countries' experiences with international capacity-building efforts.

Advocates of capacity building argue that a state with the material and organizational capabilities for carrying out important tasks is more efficient, effective, and responsive to citizens (Duit 2014; Sagar and VanDeveer 2005; Steinberg and VanDeveer 2012). Theirs is an argument in favor of using resources to build a smarter, and therefore more democratic and accountable, state, lodged against neoliberal assertions in favor of smaller state agencies, with fewer capacities (Grindle 1997, 5). In Latin America, neoliberal reforms redistributed resources and capacities from the public to the private sectors, on the theory that private entities would be more efficient, effective, and responsive than public services. In 1990s Peru, for instance, the government privatized the national electricity company and a glacier research center that monitored risks and glacier-related public safety issues. Once privatized, this research center switched its focus to provide answers to the electricity industry rather than to the state (Carey 2010). Similar shifts have been described in studies of EIA projects elsewhere, where the state takes a back seat in favor of giving industry more control (Li 2015; Goldman 2005; Silva 1997).

Part of the challenge international capacity-building efforts face is that in Chile and elsewhere, international organizations exercise contradictory influences (Hochstetler and Keck 2007). While some World Bank interventions seek to promote environmental capacity building, most encourage export-oriented industries that have a high environmental toll. Thus, capacity-building efforts for the environment focus on better governance while ignoring global market forces—driven by consumption patterns in industrialized countries—that fuel ecological destruction in Latin America and elsewhere (Sagar and VanDeveer 2005). Put simply, capacity-building efforts have not successfully countered neoliberal commitments to export-led development, global free trade (unencumbered by environmental regulation, among other things), and a small state that relies on the private sector for a large share of services.

The experience of Chile bears this out. Since the transition to democracy, political leaders and state officials have grappled with the tension between neoliberal values that privilege the market and democratic demands for a more responsive as well as accountable government (Silva 1994). These struggles play out forcefully in negotiations over how to organize science (chapter 2) and the four conflicts analyzed in this book (chapters 3–6). Across the four conflicts in this book, a steadfast collection of individual scientists, officials, and political representatives has argued for a more capable Chilean state—one able to identify and solve environmental problems in ways that are responsive to citizens. This state would require expanded enforcement capabilities, more staff, access to better technologies, more stable and generous funding sources for research, and agencies with more autonomy from the executive government. This state would also do a better job of monitoring the environment, assessing EIAs, and enforcing EIA permits. Despite some reforms—the creation of an environmental enforcement agency in 2013 stands out—these views have generally lost out to those of advocates for small government and competitive markets (Barandiarán 2016).

A second reason that international capacity-building efforts run into difficulties lies in local distrust of global interventions. In Latin American countries, the terms "international" and "global" mean "foreign" (Hochstetler and Keck 2007). Environmental interventions are particularly prone to nationalistic responses, because environmental protection is often perceived as a threat to economic growth. In many societies, foreign-led efforts appear to promote environmental protections as a thinly veiled attempt to deny the nation development opportunities (Sagar and VanDeveer 2005).[9] In parallel, other scholars and activists distrust foreign capacity-building efforts as not so thinly veiled attempts to promote foreign economic interests over those of local communities. In this view, in the guise of building capacities or protecting the environment, foreign organizations are facilitating the entry of multinational corporations interested in buying up natural assets (Li 2015; Goldman 2005; Silva 1997). The result is widespread distrust of foreign intentions that resonates with common ideas about science in Latin America, detailed in the section that follows.

Finally, there are some inherent limitations to international capacity-building efforts as relatively short-term projects carried out by transient foreign staff. Foreign-led efforts generate limited knowledge of place, in both a geographic and political-cultural sense, and frequently fail to create institutionalized spaces for dialogue or mechanisms for accountability. Many environmental issues require a longer-term perspective than short-lived efforts can provide. For instance, it took years to regulate sulfur dioxide emissions in the United States and Japan—work led by individuals deeply invested in national institutions (Keller 2009; Wilkening 2004). Intimate place-based knowledge can sometimes lead scientists to develop new conceptions of nature (Lachmund 2013). And environmental monitoring, by definition a long-term activity, is crucial to science-based protection efforts, including for the so-called baseline studies on which EIAs depend (VanDeveer 2004). But in Chile, as in other developing countries, monitoring and long-term studies are scarce (De la Maza 2001; Carruthers 2008). As the experience of CENMA illustrates, support for environmental monitoring and long-term research for regulation is erratic in Chile. For these reasons, foreign-led efforts cannot supplant the state.

Place-based knowledge is not just geographic but also social, cultural, and political. When environmental issues are involved, translating scientific knowledge into politically feasible interventions usually requires an intimate knowledge of social and political structures. Scientists with a strong knowledge of place, in a cultural sense, are in a better position to do this; they are more likely to know who has political power, how to articulate risks in ways that resonate with local residents, or which watercourses channel pollution into ecosystems that will affect local populations (Iles 2004; Steinberg 2001; Wilkening 2004). Such scientists are also more likely to identify with fellow citizens and be willing to champion the needs of communities (Carey 2010; Moon 2007). For instance, scholar Myanna Lahsen (2004) finds that Brazilian scientists felt that foreign scientists held different views on inequality, poverty, and power than those educated in Brazil. The point is not that scientists should eschew working abroad or in global scientific collaborations but instead to recognize the importance of scientists' cultural attachments to the work they do.[10]

Studies of foreign-led EIAs or capacity-building projects have highlighted the limitations of these efforts along these three dimensions: knowledge of local geography, knowledge of local society, and institutionalized accountability.[11] In an in-depth study of World Bank–led EIAs in Laos, Michael Goldman (2005) traces how global consultants lacked the time and opportunities to engage deeply with the countries where they worked, thus limiting their knowledge of local social and ecological issues. The result was poor-quality EIAs whose findings most audiences dismissed. Moreover, the scientists, consultants, and officials involved in those EIA studies had left long before local communities had an opportunity to hold them accountable for the quality and results of their work. In a study of the relationships between the World Bank, the Mexican state, and a rural community, Andrew Mathews (2011) attributes this pattern of unaccountable foreign interventions to a coping strategy used by rural state officials. The officials refuse to produce local knowledge, instead cultivating a kind of "numerical invisibility" to distance themselves from processes controlled by the World Bank and elites in Mexico City. In these and other cases, knowledge of place and accountability fell victim to the frictions between global and local expectations of the state.

A useful foil to international capacity-building efforts is the concept of "critical communities." As defined by Clark Miller (1998, 34), critical communities are social networks involved in using particular techniques and information for policy making and regulation. Critical communities can reflect on the appropriateness of specific policies, check the reliability and interpretation of scientific knowledge, articulate alternative interpretations, and expose unspoken assumptions and biases embedded in scientific claims. They are lodged in institutions, and thus can develop and sustain durable as well as tightly embedded intellectual capacities that are necessary for constructively criticizing state policies. In industrialized countries, scholars have identified critical communities with individuals and groups that exercise their critical capacities in such institutions as adversarial administrative proceedings, advisory committees, and courts run by activist judges (Miller 1998; Bjiker, Bal, and Hendriks 2009; Jasanoff 1990, 2005). The term connotes an ideal form of institutionalized governance with three salient characteristics: they are accountable, place based, and reflexive. Critical communities are an instance of and operate within broader cultures of knowing that exist in a society, or what scholar Sheila Jasanoff (2005) has called "civic epistemologies"—that is, the institutionalized relationships between individuals, groups, and organizations that sustain routine procedures used to produce, validate, disseminate, and apply knowledge to collective decisions.[12]

A new scholarship on critical communities and civic epistemologies outside of advanced liberal democracies is emerging (e.g., Bhadra 2013; Mahony 2014; Mathews 2011; Sannazzaro 2014). This body of work highlights that distrust and (sometimes violent) disagreement characterize science and public policy where powerful foreign economic and political interests meet less powerful local interests and commitments. Enlisting foreign scientists in capacity building may seem like an expedient and effective way to support sound environmental policies. Absent locally embedded critical communities, however, foreign-led interventions have rarely produced the kind of institutionalized, reflexive, and critical intellectual capacities needed for long-term environmental protections. Instead, they have led to policies that are followed in theory but not in practice, environmental injustices that are driven by unchallenged assumptions along with partial information, and locally inappropriate or misunderstood policies that are surreptitiously resisted (Carey 2010; Grindle 1997; Mathews 2011; Miller 2004; Steinberg and VanDeveer 2012; Appiah-Opoku 2005).

In the capacity-building literature like that on EIAs, the central question is under what circumstances will new policies produce substantive social change in favor of the environment. To adapt the language of early EIA advocates in the United States, When do new policies successfully "institutionalize environmental awareness"? The concept of critical communities supplies a partial answer by turning scholars' attention to the roles environmental scientists living and working in developing countries play in capacity-building efforts. Do they speak for local ecosystems and the threats they face, or for a "universal" Western science that offers standardized solutions to industrial problems? Are they accountable to local populations or foreign organizations? This book traces attempts to foster critical communities in Chile by examining the state's capacities to monitor (chapter 3), enforce rules over (chapter 4), assess (chapter 5), and study (chapter 6) the environment, either directly or through an ancillary network of knowers in universities, private consultancies, and state-owned labs.

Latin America's "Incredible" Scientists

In a 1965 treatise, Don K. Price, Harvard scholar and former adviser to President John Fitzgerald Kennedy, described scientists as the "fourth estate" of the US government. He chose the term to emphasize how deeply embedded scientific expertise was in governance. Armed with science, state officials could use numbers, data series, or graphs to evaluate the effects of a given policy, or prove that the policy had been chosen because of its anticipated effects rather than the narrow interests of a certain group. Officials could also grasp and reflect on increasingly complex issues, from the nature of warfare to chemicals to the environment, and deliver better policies as a result. It is difficult to imagine a person with political influence in Latin America ever writing a similar ode to scientists in democratic government.

Since Price's disquisition, STS scholars have produced more nuanced and accurate accounts of the power of science in governance. Science holds real and symbolic authority in advanced liberal democracies (Brown 2009; Ezrahi 1990; Hilgartner 2000; Jasanoff 1990, 2005; Porter 1995). It does so despite its well-documented practical failures: scientific claims can be used to support any number of policy alternatives, do not always provide complete or clear policy answers, and are dependent on social interests and relationships (Jasanoff and Martello 2004, 337). Science can "make environmental controversies worse" by obscuring the values, interests, and demands that different groups bring to the negotiating table (Sarewitz 2004; Kinchy 2012). Yet science continues to capture the liberal imagination—a feat achieved through any number of institutional arrangements. The influential Health Council of the Netherlands, for instance, uses advisory bodies to assert the authority of science (Bijker, Bal, and Hendriks 2009). In the United States, United Kingdom, or Germany, altogether-different routines exist to integrate science into governance in ways that protect attributes considered important for its symbolic authority (Jasanoff 2005). These include its autonomy from nonscientific interests and the closely related claim of objectivity.

Scientists' public credibility depends on their appearing simultaneously relevant to public affairs and independent or autonomous from politics (and increasingly, from business) (Brown 2009). Autonomy "refers to the condition of a collectivity that has established a social identity, a relatively stable resource base, and a system of internal social control" (Cozzens 1990, 165–166). In the case of science, autonomy is about self-governance, or the ability of scientists to hold themselves accountable to their peers versus a patron or cause. Across the industrialized world, STS scholars have documented an array of mechanisms scientists use to defend and establish their autonomy. These include organizations that can mediate between science and the state (Guston 2000) or drafting codes of conduct that differentiate "scientists" from "consultants" (Nelkin 1977). Scientists also strategically or routinely move, demarcate, or defend the boundaries of science, as opposed to nonscience, to defend their authority (Gieryn 1995). For instance, to assert their authority, US-based ecologists have differentiated science from activism (Kinchy and Kleinman 2003), while others have distinguished "basic" from "applied" science (Mirowski and Sent 2008). More recently, scholars have examined how neoliberal policies are eroding scientific autonomy in the United States (Lave, Mirowski, and Randalls 2010; Frickel and Moore 2005; Mirowski 2011; Moore et al. 2011; Berman 2012). As scientists increasingly accept industry funding, their claims to be objective are less credible because, as the adage goes, "he who pays the piper calls the tune."

In Latin America, by contrast, few, if any, elites have considered scientists members of a "fourth estate" with sway in government. Nothing like the Health Council of the Netherlands, National Academies of Science in the United States, or Royal Society in the United Kingdom—each of which provides science-based advice for governance—have ever existed in Chile or its national neighbors. The political-cultural traditions around science are therefore quite different, in particular with regard to scientific autonomy. Whereas historically in industrialized societies scientific autonomy is threatened by intrusive national governments or big business, in Latin America the threats come from three, interrelated factors: foreign meddling, resource scarcity, and cultural beliefs about scientific relevance. Understanding these different legacies is necessary for understanding ongoing struggles to define what kinds of science and society Latin American publics want as well as how they wish to participate in an asymmetrical global order.

To begin with, "autonomy" in Latin America often means autonomy from European interests or the United States. This attitude is rooted in the colonial period, when imperial powers "denied local savants and institutions full access to Western science. As a result, the little scientific work undertaken by local scientists was considered a poor imitation of metropolitan science" (Cueto 1997, 235–236). It continued during the Cold War, when the Rockefeller and Ford Foundations promoted the ideal that if US-style universities were replicated in Latin America, scientific excellence would follow (Beigel 2013; Cueto 1994; Medina, Marques, and Holmes 2014; Vessuri 1994). And it continues at the turn of the twenty-first century, when scientists increasingly participate in global research networks that offer Latin American scientists opportunity but are seen to divert their attention away from the issues that matter to Latin American societies (Lahsen 2004; Vessuri, Guedon, and Cetto 2014; Schwartzman 1987; Zabala 2010).

As scholar Hebe Vessuri (1987, 533) has put it,

> In the "First World" countries, nuclear power, genetic manipulation, and the role of "experts" are the objects of deep moral concern. In Latin America, this anxiety is subordinated to concerns about the capacity of science to satisfy the tremendous challenges of poverty, malnutrition, disease and the increasingly unequal terms of international development. Understood to be inextricably linked to economics, politics, and philosophy, science, as seen by Latin American *científicos*, appears as a spectrum of activity with no sharp edges.

If science in Latin America is an activity with "no sharp edges," as Vessuri asserts, then boundary work will be much more difficult for Latin American scientists. Scholarship suggests this is the case: Latin American scientists have struggled to define their relationship to both global research networks—are they "dependent" or "integrated"? (Kreimer 2006)—and domestic funding agencies (Barandiarán 2015b).

Latin American scientists have additionally struggled to reconcile what they see as the high standards of global science with the hard realities of local working conditions (Cueto 1997; Vessuri 1988; Schwartzman 2008; Rodríguez Medina 2014; Lahsen 2004). Latin American universities offer few full-time professorships and salaries are low, forcing many faculty members to hold multiple jobs. In Chile, these strains became so acute that scientists took to the streets in 2015 in protest, demanding wages, benefits, and full-time jobs.[13] The poverty of resources creates conflicts within scientific organizations that are difficult to repair, as illustrated by the career of Argentine Bernardo Houssay, who in 1947 received the Nobel Prize in Physiology. His career was bedeviled with fights for resources and command of the discipline. Historian Alfonso Buch (2006) argues that Houssay's fighting strategies, which included hiring foreign scientists over local ones, accentuated fractures within the scientific community, erasing any possibility of developing organizational norms grounded in a shared identity as Argentinean physiologists. Resource scarcity thus creates the conditions for "winner-takes-all" contests that weaken scientific autonomy.

The final challenge to Latin American scientists' autonomy lies in cultural beliefs about social relevance. In 1918, students occupied the University of Córdoba in Argentina to demand self-government for universities, and a professional commitment for scientists to work toward a more egalitarian and progressive society (Vaccarezza 2006). Later, during the Cold War, Marxist scientists waged an "antiscience movement" to rally science as a tool to gain the region's political, social, and economic autonomy from conservative local elites and the United States (Herrera 1974; Sábato 2004). These demands for a socially engaged science challenged mainstream practices that typically focused on narrow scientific problems, defined by the interests of First World scientists, or favorable niches, like high-altitude medicine (Benchimol 1999; Cueto 1989; Diaz, Texera, and Vessuri 1983; Kreimer 2006; Kreimer and Lugones 2002; Roblero 1995; Vessuri 2007; Zabala 2010). A legacy of Marxism in Latin America is thus a general distrust of scientific autonomy, as the concept implies that scientists are working on issues defined by a scientific agenda as opposed to issues of relevance to Latin American society (Vessuri 1987; Beigel 2013). Whereas in industrialized societies scientific autonomy, understood as scientific self-regulation, is celebrated, some sectors in Latin America view it as selfish and elitist.

These decades-long debates reveal ongoing struggles to orient science toward a certain vision of society—one dedicated to fighting poverty and inequality, against the power of local elites and their allies in the United States and Europe. From a coproductionist perspective, the making of the state, society, and science are locked in bitter negotiations, made more stressful because, to paraphrase Vessuri, the boundaries of science are blurry. The nature and durability of these debates has consequences for the role of science in a democracy. To the extent citizens see their national governments as aligned with global or foreign interests, scientists who participate in capacity-building efforts, or attempt to build an embedded critical community, are likely to face criticism from those convinced that neither the government nor foreign entities are committed to addressing local problems of poverty, inequality, and injustice. Scientists who participate in global scientific collaborations that bring resources, equipment, and scientific prestige become similarly suspect. Moreover, the structure of these collaborations only highlights Latin American scientists' low social status—both national governments and international organizations reproduce a hierarchy between First World and Latin American science that erodes Latin American scientists' social status within their own societies (Lahsen 2004; Rodríguez Medina 2014; Shrum and Shenav 1995). Those who practice a science with no sharp edges have few tools to demarcate its boundaries to defend its authority.

The chapters that follow describe these dynamics for Chilean environmental scientists trying to participate in government through EIAs. The conflicts analyzed illustrate how, in responding to the opportunities around them, Chilean scientists working at universities, research centers, and state-owned labs exposed themselves to constant skepticism. Their possible patrons in industry, NGOs, and the state suspected that scientists lack an independent voice. They are, moreover, opportunistic, parochial, and rudimentary—in Chilean slang, scientists study "the immortality of crabs" with age-old methods rather than publicly important issues in sophisticated ways. In response, Chilean scientists used two surprising strategies to defend their credibility. They reached results contrary to their funders' interests and narrowed science down to a few tasks of seemingly unassailable objectivity, like keeping an inventory of the landscape.

Absent a locally embedded and credible critical community, the patrons of scientific advice in Chile often preferred to hire foreign scientists, who benefit from the assumption that they are independent of financial and political interests in Chile. But these foreign scientists too are suspect, as their presence draws into question Chile's autonomy from foreign agendas. Their critics point out that the asymmetry in foreign and domestic funding, equipment, and training necessarily creates a hierarchy with foreign science at the top and local science at the bottom. Far from empowering their local colleagues, foreign scientists unwittingly disenfranchised their Chilean peers by offering their own solutions that shut down, marginalized, or discredited those offered by locals. Like a deus ex machina in a play, the arrival of foreign scientists might bring closure to a specific policy debate, but does not contribute to building durable, institutionalized, reflexive capacities. Their arrival thus perpetuates an intellectual dependence that has mired Latin American societies in bitter debates about science and society, and weakens efforts to build effective critical communities that can strengthen the quality, pluralism, and accountability of democratic government.

The Nuts and Bolts of EIAs in Chile

EIAs worldwide follow a template within which practices vary (Glasson, Therivel, and Chadwick 2012). As background for what follows in the rest of the book, I here describe the basic outlines of Chile's EIA process. When a company or other entity wants to build a project that will impact the environment in one of several ways listed in the regulatory decree that implements EIAs (Decree 95, article 11), it is required to conduct an assessment of those impacts. Based on these reports, state officials issue an EIA permit that can, in theory at least, be suspended in the event of violations.

EIAs typically consist of two parts: a baseline study and an impact assessment. Baseline studies record existing ecological conditions, while impact evaluations assess and rank how the project will impact those conditions. The company pays for this work, typically hiring consultants and occasionally scientists. The impact assessment usually consists of ranking hundreds of potential impacts, on topics from water quality to solid waste management, on a zero to one hundred scale. For each negative impact the EIA diagnoses, the SEIA verifies whether the project meets applicable laws, and proposes measures the company should take to eliminate, reduce, or compensate for each impact. The SEIA then proposes under what conditions the project can be approved, and a committee of ministerial representatives votes to approve or reject the project.[14] The committee can withhold approval if the project violates environmental quality standards or is expected to impose undue burdens on the affected communities. In Chile and elsewhere, authorities have rejected around 5 percent of EIA projects.[15] In practice, governments primarily use EIAs to improve industrial projects, not reject them (Glasson, Therivel, and Chadwick 2012).

In Chile, EIA can be used to refer to a policy, a process that applies regulation through an assessment, or an administrative entity staffed by state officials who assess EIAs (I use the singular EIA to refer to a specific study). Between 1994 and 2010, EIAs were administered by the SEIA, which operated within Conama. On its creation in 1994, Conama was intended to be a small coordinating agency that would encourage production-minded ministries, like mining or agriculture, to incorporate environmental concerns into their work. In addition to managing EIAs, Conama was responsible for recommending pollution control and environmental quality standards, but the agency never had the personnel, resources, or authority to fulfill these roles (OECD 2005). In 2010, Conama was replaced by the autonomous EIA Agency. Throughout the book, I will refer to both Conama and the EIA Agency, as these were the two organizations responsible for EIAs in Chile—the first between 1994 and 2010, and the latter since 2010.

Neither Conama nor the new EIA Agency and Ministry of the Environment have in-house scientific expertise. Instead, they rely on ancillary networks of knowers that include personnel working at universities across the country, research centers that operate autonomously from universities (e.g., the Center for Oceanographic Research in the Eastern South Pacific [COPAS], Center for Advanced Studies in Arid Zones [CEAZA], and Center for Ecology Research in Patagonia [CIEP]), a few state-owned research centers (e.g., CENMA, the Fisheries Research Institute [IFOP]), and for-profit consulting companies (e.g., Gestión Ambiental Consultores [GAC], Arcadis, Golder Associates, and Centro de Estudios Ambientales [CEA]). These organizations provide information to Conama and other state agencies through the EIA process, and staff members in each agency—a few with relevant technical training—evaluate that information. Most of the information included in an EIA is produced through the potential licensee, who hires scientists and consultants to complete baseline and impact evaluation studies, and then answer questions the state agencies might have.[16]

Conama has 120 workdays to assess an EIA. It first forwards the EIA documents to each supporting agency for review and then organizes public participation meetings for affected residents. It then uses this information to request clarification or improvements from the company. The 120-day clock stops only when the company is formulating its responses to Conama's requests.

The EIA documents (baseline plus impact assessment reports) become public information available to anyone, worldwide, through an online database of EIA projects. The company retains ownership, however, over the scientific studies conducted for the baseline that consultants decide, for whatever reason, to exclude from the EIA documents (for examples, see chapters 4–5).

In his classic account of scientific advice within the US government, Price (1965) contrasted "scientists," who are not purpose driven, to "professionals," who translate science into actionable goals. "Administrators" in his model are public employees who, in theory, answer to citizens, while "politicians" hold the most decision-making autonomy. These distinctions help to map the range of functions different individuals and groups play in Chile's EIA. Some of these roles are legally defined; for instance, administrators are either civil servants or political appointees employed by state agencies involved in EIAs. But others, like those of scientists and consultants, are subject to boundary work as individuals struggle to participate in EIAs while defending their credibility and that of the EIA process itself. As the next chapters show, this is a challenging task, in large part because Chile's environmental institutions and EIAs provide local actors with a circumscribed role. What counts as a significant environmental impact is defined not by state officials or scientific methods but rather by law. Companies, following the rules and regulations of EIAs, decide what to submit for baseline studies. State agencies have few resources and little political authority to challenge them. As the rest of this book will make clear, absent critical communities, the mythical Chilean way cannot address the myriad concerns raised by EIA studies.

2   The Social Contract for Science in Chile

During Pinochet's military dictatorship, scientists gathered to talk about the place of environmental science in Chilean government. Public meetings were rare in those days. But in 1983, a civil society organization called the Center for Environmental Research and Planning (Centro de Investigación y Planificación del Medio Ambiente, or CIPMA) convened a town hall–style meeting to discuss how science might better serve society through the environment. Interest was staggering; three hundred people from academia, civil society, and business attended that first meeting, and five hundred came to the second meeting held three years later. To this crowd of five hundred, biologist Ernst Hajek (1987, 13) asked, "Often scientists are accused of being stuck in their ivory towers, facing the academy and with their backs to the country's real problems. ... But scientists could also ask, and how do you enter that other ivory tower, the organizations where decision-makers work?" Hajek thus turned the common trope—that scientists work in ivory towers—on the government. Marginalized from government, how could scientists propose solutions to the country's environmental problems? When democracy returned, as everyone expected it to, would scientists continue to be sidelined, or would they recover or perhaps expand the largely progressive role they played before the coup?

This chapter traces the relationship between Chilean science and the state during the twentieth century, paying particular attention to shifts in what can be called the "social contract" for science. Scholars have described a social contract between scientists and the US government, in which the government agreed to massively fund science while respecting scientific autonomy, in exchange for scientists agreeing to provide the state with strategic advice and technological advances (Guston 2000). As a theoretical device, the social contract is useful for exploring the norms guiding the organization of science over time and highlights this book's coproductionist argument (Jasanoff 2004). Following the fates of scientists and the organizations where they worked—primarily universities—offers insights into how positivism, Marxism, reactionary conservatism, and neoliberalism have influenced Chilean society and government over the twentieth century.

The narrative is organized around four shifts in the social contract for science. For decades before the military coup, Chile was governed by a relatively stable liberal democracy that sometimes promoted scientific forms of governance to modernize the country and at other times held science back for fear of secularizing society. During the 1950s and 1960s, Marxist ideas challenged these debates, calling on scientists to devote themselves to finding solutions to the needs of the poorest members of society. This was a time of tremendous growth in science at universities and a handful of state-owned research centers were created to supply the state with strategic advice. This activity was violently interrupted in 1973. After the military coup, a decade of repression of university students, staff members, and professors followed, succeeded by a decade of neoliberal reforms and, after the transition to democracy in 1990, the consolidation of those reforms. The military government's policies were characterized by two goals. First, to eradicate Marxist ideas that had encouraged universities and scientists to pursue work with a "public vocation"—that is, oriented toward socially progressive, pro-poor goals. Second, to create more opportunities for the private sector, including in higher education and science. After 1990, under Chile's new democratic leaders, scientific advice ceased to be "strategic" and was increasingly traded via a market.

Through this period and debates, the scientific community was far from monolithic. Some scientists objected to Marxist ideals and demands. Many (though not all) resisted the military government as best they could. CIPMA meetings in particular were a site of such resistance: a space in which participants could articulate their hopes for a democratic future when environmental science might provide scientists with opportunities to serve society as government advisers on this important issue. Some scientists have also protested the expansion of market mechanisms into science, concerned that the market would erode scientific quality and credibility while failing to provide the state with necessary advice and information.

One useful way to study these disagreements is through boundary work undertaken by different actors to differentiate science from nonscience.

Chilean scientists and their patrons have disagreed about how to demarcate scientists from consultants and basic from applied science. At stake is who has the privilege to be called "independent" and "academic"—two terms that in the US social contract for science, for example, underpin the special authority of scientists in government (Guston 2000): as members of academia, scientists set their agenda according to the internal norms and demands of science and their peers. This makes them independent—a quality that advocates of the social contract claim is further guaranteed by state sponsorship for science.[1] But in Chile, as this chapter shows, the meaning and uses of these terms are far less clear as a result of low scientific autonomy, low public support for science combined with violent hostility against universities, and the relatively recent rise of neoliberal values.

Around 1990, the future looked bright for CIPMA scientists. New environmental policies, organizations, and labs seemed to promise opportunities for becoming involved in environmental governance. Yet by 2010, these felt inadequate to scientists: none were providing the government with scientific advice. Nor did Chile's democratic leaders seem concerned; as the twenty-first century started, the dominant view among them was that the state could forgo environmental scientific advice. The first part of the chapter traces Chilean science before and through the military coup of 1973. The chapter then examines the transition period and organization of environmental science post-1990. Two new labs created at this time, CENMA and the Europe-Latin America Center (EULA), might have fulfilled the hopes of CIPMA scientists. The final section of this chapter analyzes the trajectory of these two labs to pinpoint specific institutional arrangements that have enacted neoliberal values in Chilean environmental science and governance.

Science before 1973

In the mid-nineteenth century, the new Chilean state turned to science to help assert its identity and power, and project these to other nations. Following the examples set by Britain and France, which sent scientists around the world with military expeditions, Chile's Congress sent natural scientists to study the new nation's geography and inventory its natural riches (Jaksic 2013). For instance, French natural scientist Claudio Gay explored central Chile, parts of the Atacama Desert, and other regions, sending back descriptions that "demonstrated the endless national potential to Chileans, while linking nature with Chilean national identity" (Schell 2013, 44). As the twentieth century approached, Chilean science was thriving: the University of Chile trained scientists and engineers, and the Chilean National Museum of Natural History was internationally and domestically renowned. "'Scientific' had become a positive adjective meaning modern, rigorous, and up-to-date" (ibid., 204).

Also around this time, individuals like Valentin Letelier and Jose V. Lastarria added to the lure of science by promoting a "scientific" government. Letelier and Lastarria identified as positivists, and echoed global progressive principles in which public decisions should be based on reason and expertise to defeat the interests of the oligarchy. Positivist ideals led to the creation of the Production Development Corporation (Corporación de Fomento de la Producción, or Corfo) to promote the state's scientific view in economic development (Silva 2009) and to the creation of a teacher training college, the *Instituto Pedagógico*, which until the coup trained or employed some of Chile's greatest minds (Jaksic 1989). After World War II, international funding for this kind of progressive scientific capacity increased. Several state-run research institutes were built, with expertise in soils, fisheries, forestry, and other things (Bustos 2015; Illanes 1993; Klubock 2014; Tinsman 2014; Venezian 1987).

During the 1960s, Chilean universities grew to attract scholars from across the region. Inspired by the Cuban Revolution, many Latin American scientists began to advocate for a purpose-driven science that could solve "Third World" problems of poverty, disease, and inequitable growth. Foreign funding became especially divisive, as some scientists rejected it as an instrument of US anti-Communist foreign policy. In Chile, however, many embraced foreign funding, which through the 1960s helped expand opportunities for science and intellectual activity in the country (Beigel 2013). The Chilean state also played a key role by integrating scientists and foreign aid into governance, to support socially progressive policies in agriculture (Tinsman 2014), health (Illanes 1993), and economic planning (Medina 2011). For ecologists, the Instituto Pedagógico stood at the center of this "golden age for culture in Chile" (Montecino and Orlando 2015, 216; Jaksic, Camus, and Castro 2012).

Achievements aside, twentieth-century Chilean science also faced major challenges that collectively reinforced the scientific community's lack of autonomy. One problem involved the colonial roots of Chilean science.

Scientists continued to look to Europe and later the United States for recognition. Gay, for instance, eventually returned to live in Europe and took most of his specimen collection with him. The first director of the Chilean National Museum of Natural History, Rodolfo Philippi, anguished over Europeans' reviews of his museum rather than celebrating the local popular acclaim he enjoyed (Schell 2013). As scientists turned to Europe and away from the state that funded them, their funding was never constant or assured (González Leiva 2007; Villalobos 1990). For example, within a few years of reaching its height of world acclaim, the Chilean National Museum of Natural History lost public support and fell into permanent decline.[2]

Second, rivalries and conflicts crippled many scientists' careers. The Chilean state, when controlled by conservative leaders, withheld funding for science, which conservatives saw as a secularizing force. This fear motivated them to create the Catholic University of Chile (Jaksic 1989). For most of the twentieth-century, the Catholic University and state-run University of Chile were ideological rivals and competed for funding and scientists. Arguably, this competition has not strengthened the autonomy of Chilean science. The career of Hector Croxatto, Chile's most accomplished research scientist, helps to illustrate this. Croxatto spent years shuffling between the two universities, fielding accusations that he was misspending public funds—his laboratory consumed too many resources in pursuit of a narrow goal—or had unfairly criticized the state university (Roblero 1995). These rivalries were not just about resources but rather about scientists' social standing: Was their research useful and relevant to society or, as Croxatto was accused of, was it focused on "narrow" scientific goals that might never benefit Chileans?[3] In the context of the Cold War, this question was an instance of broader debates that pitted Marxists against the rest (Beigel 2013). A result of these debates is that scholars, scientists, and others are frequently suspicious of scientific autonomy; the thinking is that autonomous scientists turn their backs on society (Kreimer 2006; Vessuri 1987). By contrast, in the US social contract for science, the opposite is true: without scientific autonomy, scientists' credibility as scientists is suspect (Jasanoff 1990).

Finally, Chilean scientists have typically been resource poor. State funding for science has historically been low and erratic, as it continues to be since 1990 (Brunner 1993; Allende et al. 2005; OECD 2009). For instance, Chile's flagship natural science journal, the *Revista Chilena de Historia*

Natural, kept going thanks to the personal funds of its chief editor (Jaksic, Camus, and Castro 2012). Research funding was often extremely low. At the University of Chile, for example, Croxatto could not afford to buy the animals he needed for his research on kidneys, and his salary was barely sufficient to rent a small house in central Santiago. Forced to work multiple jobs, he had to reduce his time in the laboratory and spend more time teaching, including training science teachers at the Instituto Pedagógico—a position for which he almost paid dearly (Roblero 1995). After the coup, the military government and its supporters used Croxatto's association with the Instituto Pedagógico to rebuke him because, in their view, the institute represented the worst of state intervention in public life (Quesada 2013). In short, like other Latin American scientists (Cueto 1997), Croxatto had to fight for scarce resources, defend his work from accusations that it was irrelevant to Chile, and when he left the lab to contribute to society as a teacher, step into a destructive political process that nearly cost him his life.

Repression and Reform, 1973–1983

On September 11, 1973, Chile's military staged a coup to depose social-ist President Salvador Allende, and eradicate what they saw as radical and Communist persons from society. The military hit universities hard. It detained thousands of students, professors, and administrators, sending them to their deaths or exile (Constable and Valenzuela 1991). During the final months of 1973, army generals replaced scholars at the head of every university in the country. The University of Chile was dismantled; its ten campuses and the Instituto Pedagógico were either closed or separated by decree from the main campus (Brunner 1986; Levy 1986). Immediately after the coup, approximately 20,000 students and staff members were forced to leave the University of Chile, and over the next decade, seats for students were cut from 146,000 spots before the coup to 118,000 afterward. The University of Chile's intellectual workforce fled, and ecologists moved to the Austral and Catholic Universities (Jaksic, Camus, and Castro 2012). The expansion of opportunities for university-based study and science made during the 1960s were thus painfully reversed, and the former University of Chile system—formerly Chile's flagship with campuses nationwide—fell into permanent financial hardship (Constable and Valenzuela 1991; Monckeberg 2007).

Repression was followed by neoliberal reforms that changed who could attend university, what they could study, and professors' and scientists' work. The goal of these policies was to permanently immobilize universities with a public vocation (Constable and Valenzuela 1991; Jaksic 1989; Monckeberg 2007). Policies introduced in 1981 cut public funding for universities, introduced competition for students, and made it easy to set up new private universities (Brunner 1993). Life inside universities during the 1980s was difficult: a rigid hierarchy existed, and creativity and productivity suffered (Jaksic 1989; Levy 1986). Scholars turned to "safe" spaces for research, such as small institutes sponsored by the Catholic Church and foreign donors that employed faculty forced to leave universities. Those able to retain their university position turned to "safe" research topics that avoided social questions or the needs of the poor and vulnerable. Thus, geographers abandoned human geography to study rivers and mountains, historians focused on colonial times, and so on (Jaksic 1989).

Interestingly, the environment was considered a safe area of research (Nef 1995). A founding member of CIPMA recalled that by portraying themselves as "just environmental scientists," they obtained from the military government permission to travel to conferences abroad and organize their large meetings.[4] That the military government tolerated environmental science is perhaps not surprising; for most of the twentieth century, state agencies used environmentalist arguments to consolidate their control over Chile's southern forests (Klubock 2014). Scientists who took up environmental causes in the 1980s—such as the capital's growing smog problem or conserving Chungará Lake from mining in the Atacama Desert—did not question the government's development model (Camus and Hajek 1998). Pinochet's administration even signed a few environmental treaties and created a small ecology unit under the direction of Juan Grau, a respected doctor (Jaksic, Camus, and Castro 2012). What is surprising, then, is how quickly and forcefully environmentalism penetrated social movements; by the 2000s, demands for environmental justice were central to a range of social movements (Klubock 2014).

Scientists' Hopes for Democracy

CIPMA and the meetings it held in the 1980s were one such safe space in the dictatorship years. At the meetings, scientists could reassert the public vocation that military operatives had tried to stamp out and stage a well-behaved resistance—the kind the military government might tolerate without inflicting further physical harm. At a CIPMA meeting, an elderly Croxatto commemorated the passing of his old friend Juan Gómez Millas, who in a fifty-year career as faculty, education minister, and chancellor of the University of Chile, fought to expand access to higher education and helped make the Instituto Pedagógico as well as the University of Chile into the organizations the military regime so fiercely hated. Croxatto (1987, 24) recalled, "This concern of his spirit, moved by his desire to educate, instinctively led Gómez Millas to direct his attention towards these crusades championed by CIPMA, of interdisciplinary studies, to…reflect on how to promote actions that will lead to a long-term balance between humans and nature." Under dictatorship, scientists could no longer champion social change as they once had, but they could now be environmentalists.[5]

In the pursuit of new opportunities for scientists to contribute to the betterment of society, science provided CIPMA participants with a measure of protection during the dictatorship. Understood as an act of boundary work, this speaks to the attributes of science at the time as politically isolated and unthreatening. The CIPMA events were clearly targeted at academic scientists. The meetings were sponsored by Chile's Academy of Science, and speakers addressed the crowd as academics and scientists (e.g., Hajek 1987). Speakers described the responsibilities of scientists in ways that echoed liberal ideals in which science delivers objective knowledge to help solve social problems. Statements like "scientists are expected to deliver an objective diagnosis of the key problems and ways to solve them" (Arenas 1988, 78) were common. Similarly, participants suggested scientists should identify a portfolio of sustainable projects that industry should undertake. They proposed to define "quality of life" and "sustainability" so as to set a standard for industry to achieve. Others called on scientists to develop better theories and models of environmental change. These proposals echoed the founding values of NEPA, the legislation passed nearly fifteen years earlier to connect scientific advice to environmental decision-making, which CIPMA participants held as a model to follow. Fundamentally, they hoped an interdisciplinary and objective science would yield concrete solutions to real environmental problems.

Scientists at these meetings saw government and state officials as their primary interlocutors. For example, Hajek (1987), who criticized the government for being itself an "ivory tower," lamented that "traditionally, few people with technical training work at decision levels [in government] where they can formulate demands on the scientific community." The solution, he suggested, was for individuals with scientific training to occupy more positions in state agencies and government. A few scientists also deplored the "disconnect" and "distrust" between science and industry, while students at the meetings felt strongly that scientists needed to be in closer communication with activists.[6]

The third CIPMA meeting, held in 1989, was marked by optimism. CIPMA participants hoped that the new democratic government—then already discernible—would inaugurate institutions, regulations, and policies that would give scientists a role in environmental protection. Of these, EIAs seemed particularly promising.[7] In 1994, for example, the country's most renowned ecologist, Mary Kalin, led the studies for the Trillium forestry project, the first high-profile EIA done in Chile. This became a difficult learning experience, however; the scientists set up an Independent Scientific Commission, and signed a charter with the company's chair to protect their autonomy to speak to the press and recommend whatever they saw fit (Kalin et al. 1996). Nevertheless, their peers criticized them for working with industry, and the project was eventually terminated due to environmental concerns (Taylor 1998). Although EIAs seemed to offer opportunities for scientists to participate in governance, the Trillium experience soon showed the difficulties involved in finding acceptable terms that would allow scientists to navigate suspicions that they lacked autonomy or had "sold out" to industry.

Another promising trend for scientists involved the creation of new environmental science institutes: EULA and CENMA, as of this writing, are still Chile's only research institutes focused on the environment. EULA came out of aid from Italian universities that approached the University of Concepción to initiate research into the water quality and hydrodynamics of the mighty Bío-Bío River. The Italians were commemorating five hundred years of Christopher Columbus's voyage to the New World, and looked to support science so as to help "Chile think about Chile" (Gyhra 1989). Among other things, EULA developed a PhD program in environmental sciences, the first and still only of its kind in Chile.

Around the same time, CENMA was set up to provide Conama, the environmental state agency created in 1994, with environmental information and advice. As the transition began, a group of CIPMA engineers contacted JICA for funding support. JICA responded enthusiastically; it saw an opportunity in Chile to replicate Japan's approach to scientific advice, where federal and regional environmental agencies have in-house scientific expertise. In 1995, JICA committed US$20 million to build modern air and water quality laboratories, employ dozens of full-time scientists, and support a generous exchange program. CENMA initially provided Conama with important services that built the agency's intellectual and assessment capacities. In particular, CENMA (1999, 2014) trained dozens of state officials in how to do EIAs.[8]

Democracy Is Restored, but Not Science

When presenting the environmental framework law that created Conama and required EIAs to the Senate, President Aylwin had some words of recognition for Chilean scientists: "From the beginning of our history there have been in Chile men and women who have alerted us of the fragility of our land and who protected our natural resources. Many worked in silence, often misunderstood and with little support, studying the resources of the nation and deciphering her riches" (LH 19.300, 9).[9] As the Senate debated the environmental framework law, legislators faced three concrete decisions where they might have granted scientists an institutionalized voice in environmental governance. Scientists might have gained a special seat on Conama's Advisory Council, been hired as Conama staff, or benefited from earmarked research funds administered by Conama. Yet senators perpetuated the trends observed by President Aylwin. Environmental scientists continued to work "misunderstood" and "with little support." Legislators demonstrated a general skepticism of science as well as a neoliberal attachment to competitive markets as the best way to organize society. Together, these values underpinned specific choices that have brought the neutral broker state into being.

Conama's Advisory Council aimed to give civil society, broadly construed, a say in Conama's business. Between 1994 and 2010, the council reviewed recommended environmental quality and emission standards, and heard appeals against EIA decisions. It included two members each from academia, NGOs, business, labor, and consulting companies, and one representative of the president. The council was thus set up to be like a Noah's Ark that "balances out different estates of society" (LH 19.300, 53, 205).

For scientists, the definition of who would participate on the Advisory Council was a key moment of boundary work. As Conama's only advisory body, legislators could have given science a special voice in environmental affairs, but instead granted scientists two representatives like everyone else. Any special status scientists might have had was further eroded by the fact that senators also gave consultants two seats at the table. Consultants were not included in the original proposal. A legislator added them, calling them "private entities dedicated to research on environmental issues," to refer to for-profit companies hired by potential EIA licensees to prepare EIA studies and documents. Under the proposal that the Senate ultimately approved, consulting companies like GAC and ARCADIS have selected individuals from among their employees to sit on the council. By contrast, the Association of Research Universities selected scientists from among tenured faculty.

With this decision, legislators demarcated consultants and scientists, giving them equal and separate voices on Conama's Advisory Council. But they then defined consultants in such a way that they also blurred the social or cultural differences between them. During the legislative debate, senators redefined consultants from private entities to *"independent academic* centers that study or work on environmental issues" (LH 19.300, 435, 500–507, 923; Law 19.300, article 78(c), emphasis added). The legislative record does not clarify what motivated this significant change. Independent and academic are not random words; in the United States, for instance, these traits underpin scientists' special authority as objective parties in government (e.g., Guston 2000; Hilgartner 2000; Jasanoff 1990). In the Chilean case, this change suggests that at least among the country's senators, independent means autonomous from the public versus the private sector, and academic applies to commercial science that is not published in peer-reviewed journals. In practice, over the next twenty years, consulting companies participated enthusiastically on the council. By contrast, scientists generally saw it as a "representative," not an "expert," forum that was important in principle but not in practice.[10]

Another lost opportunity for scientists came with regard to loosening the requirements to qualify for employment at Conama. Legislators rejected a proposal to allow Conama to hire individuals with a PhD or masters-level training in an environmental field, but who lacked a professional degree. In Chile, as in other Spanish law countries, some bachelor degrees include a professional degree. This proposal would have facilitated hiring people trained in countries like the United States where bachelor's degrees do not enable one to exercise a profession. Legislators rejected this proposal because some senators were concerned about quality; one legislator asked rhetorically, "If tomorrow someone appears with a degree from the University of Paramount, Vernon North or whatever else, who would be able to evaluate that?" (LH 19.300, 668). Admittedly, this decision likely had little effect on who worked at Conama because most Chileans get professional degrees. Still, it is interesting because this is the only moment in the legislative record when legislators showed a concern with academic quality. Conversely, they largely ignored the problem the proposal had sought to solve: how to train the nation's environmental workforce and ensure that Conama could hire staff with specialized knowledge.

The final missed opportunity involved the creation of a special fund for environmental science. Some senators argued passionately for more research, for instance, to inventory the nation's environmental problems, identify sites that should be protected, and propose sustainable projects worthy of public and private support. To meet these needs, one legislative proposal suggested that the National Science Agency (Comisión Nacional de Investigación Científica y Tecnológica, or Conicyt) run a special program for environmental research. Another suggested creating an Institute of Environmental Quality to strengthen the state's research capacity. But these proposals were rejected due to the dominance of neoliberal values that have fostered a neutral broker state in Chile. Legislators who opposed earmarking public funds for environmental science argued this would have contravened market competition. In this view, environmental science should compete for funding on a "level playing field" against health, welfare, and other issues (LH 19.300, 1002–1013). Then Senator Piñera (later president) summarized this viewpoint when he said that the nation's general budget was the only special fund the environment should draw from. In other words, the Senate had no business prioritizing issues or earmarking funds for acute needs. Other legislators wanted to leave everything to consultants; from their perspective, EIA studies done by consultants would produce all the environmental information regulators needed.

These neoliberal views largely prevailed, although some compromise was necessary as not all legislators agreed. Thus, Senators approved a "fund for environmental protection," but it has had a tiny budget—only enough for two to three small projects per region per year. When it came to consultants, however, the Senate placed no constraints on them, despite some dissenting voices that anticipated the problems with credibility and conflicts of interest cataloged in this book. After EIAs became required, consultants became central to EIAs. Their economic power grew, and they developed a close relationship with Conama; individuals move regularly from Conama to consultancies and vice versa.[11] Some former Conama employees complained this has placed blinders on Conama's expertise, as most consultants (and by extension, Conama officials) train in environmental engineering as opposed to specialized areas like soil science or freshwater biology.[12] Consultants have, moreover, cultivated special relationships with political parties and thereby gained political power. The company GAC was close to Piñera's government, and ARCADIS has traditionally been close to the Concertación parties.[13] As a result, consultants—not scientists—have gained a privileged voice in environmental policy. For example, consultants dominated an advisory body set up in 2016 to recommend reforms to EIAs (Propuesta Comisión 2016).

Undoing CENMA

Another effort to create environmental scientific capacity for the state, specifically for the new agency Conama, was the creation of CENMA. This was supposed to be owned and operated by Conama to provide the agency with privileged information along with advice on everything from environmental quality standards to issues related to specific industrial projects. The initiative came from Chilean engineers who, based on their experiences studying air quality in Santiago, felt that a democratic government was going to need scientific capacity. As noted earlier, this vision resonated with JICA, which saw an opportunity to replicate Japan's model of scientific advice in Chile. The engineers and JICA, though, disagreed on how best to organize scientific advice in Chile. While JICA was committed to a state-centric model in which a state agency like Conama would have internal expertise, the engineers believed this model was "unmanageable" in Chile. Their arguments reflected neoliberal values: they asserted that by law, the state cannot have a privileged relationship with any one group. In-house science risked bloating the state and violated free market principles by creating an unfair advantage for CENMA (JICA 2002).[14]

The alternative to an in-house lab for Conama was to constitute CENMA as a private foundation on lands ceded by the University of Chile. This arrangement worked while CENMA had JICA funding. CENMA built world-class laboratories and trained dozens of scientists as well as hundreds of state officials who worked on EIAs. JICA extended its support for two years so as to secure its legacy against what it saw as lackluster support from the Chilean state.[15] JICA's fears were soon confirmed; just three months after the 2000 agreement was signed, Conama cut its contribution to CENMA by two-thirds.[16] After JICA funding ran out, neither Conama nor the University of Chile have consistently funded CENMA, and its budget has been volatile as a result. It fell to less than US$1 million a year, then tripled after Conama signed a multiyear agreement for services from CENMA (2008, 2009, 2010). In 2011, the budget again fell dramatically, and dozens of scientists left. In 2017, CENMA remained in limbo, slowly losing its talent and seeing its equipment fall into disrepair.[17] The second floor of CENMA's offices has remained closed due to structural damages to the building caused by the 2010 earthquake, and has yet to be repaired.

Barred from becoming a government laboratory, CENMA has had to continually reinvent itself, prompting scientists, professionals, administrators, and politicians to wonder what exactly the center is, and why they should trust it. The first tension that plagues CENMA is that between basic and applied science. Without funding from JICA, CENMA's work has oscillated between what many called "basic" and "applied" science.[18] In this context, the difference between these has to do with scientific autonomy: basic science answers to the interests of scientists at the University of Chile, while applied science answers to Conama's and other regulators' needs. The chancellor of the University of Chile appoints CENMA's director from among tenured faculty. CENMA's first director pursued CENMA's applied mission but was succeeded by a director more interested in basic science.[19] Forced to veer between basic and applied science, CENMA's reputation has suffered. Many felt that the lack of continuity led CENMA to do neither kind of science well, while others expressed little sympathy for CENMA's original mission. In their view, applied science is not interesting compared to basic science.

In a bid to increase its revenues, CENMA attempted to reinvent itself as a national reference center. This began as a metaphoric goal that resonated with JICA's original vision: CENMA would be the nation's intellectual center of reference on all things environment. But as CENMA's finances worsened, this also became a literal mission. With aid from Canada's environmental agency, CENMA purchased equipment and learned the operating procedures needed to obtain ISO 17.025 certification, required by the global mining industry. ISO 17.025 certifies that a lab is competent to test and calibrate equipment as well as materials, and ensures the traceability and comparability of data. CENMA's leadership hoped this would prove CENMA's value to society and generate revenue through the sale of analytic services. They were soon disappointed. Because Chilean regulation does not require ISO 17.025, mining companies were unwilling to pay a premium for laboratory work that met international standards.[20] This expensive and time-consuming effort thus ultimately failed.

As a last resort, CENMA began to compete against consultants and university-based scientists for industry and government contracts, the latter obtained through Chile's *mercado público*, or public market, used by government and state agencies to purchase everything from pencils to scientific advice. CENMA's experience with one contract in particular supplies evidence of how far the Chilean state is willing to take free market competition and how unaware officials are of the value of scientific data for environmental governance.

Since the 1990s, the Ministry of Health has hired CENMA to operate air quality monitoring stations in several cities. The ministry used these data to issue public health warnings that triggered restrictions on outdoor, vehicle, and industrial activities. After nearly twenty years, CENMA lost the tender to a small start-up. The Ministry of Health's decision was based on price: it had cut the budget for air monitoring by a factor of three, and CENMA's proposal was 11 percent more expensive than the competition. In the equation the Ministry of Health used to rank bids (standard in Chilean government), experience counted for 30 percent and cost for 25 percent. Applying the equation to the competing proposals, CENMA earned just 0.3 points (out of 6) for its two-decades worth of experience, while the cheaper start-up—which had just three employees and zero experience—got more points and won the contract.[21] This decision terminated a data series with over ten-years of air pollutant measurements and CENMA's air quality laboratory—in JICA's days, considered one of the crown jewels—closed from lack of work. Skilled and experienced scientists who had joined CENMA to protect citizen's quality of life were therefore forced into consulting work.

The ministry's decision hurt citizens' health too. A study of the effects of this switch found that, for example, in the city of Rancagua, the authorities did not have sufficient data 84 percent of the time to assess whether or not an air quality emergency existed (SustenTank 2011).[22] Because half the days in a four-month period registered invalid or missing data, the authorities could not issue air quality warnings. This study also points to the disarray that exists in the state's air monitoring network. Nationwide, of the 160 installed air quality monitoring stations, just 40 belong to the state and the rest to private companies. As a result, some areas are overmonitored, others are undermonitored, and there is no benchmark data series against which to measure the effectiveness of new air quality or pollution control policies. Instead, the government responds to crises when they happen; for instance, after the residents of the small town of Andacollo protested because they refused to believe air quality data collected by the local mining company, the Ministry of Health installed its own monitoring station there—the fifth in this small town.

Many blame the University of Chile for CENMA's troubles. They argue that the university has looted CENMA for resources and successive directors ran the lab for their personal projects. Given the history since 1973 of lackluster support for public science and outright hostility against the University of Chile, however, both CENMA and the university are victims of government indifference or aversion to scientists with a public vocation. A few remaining longtime CENMA employees recall that in its heyday, CENMA was a "stone in [the government's] shoe." For example, the lab would directly report an air quality emergency to the press that the government would have preferred to call a simple "warning."[23] Another employee recalled efforts to offer the government scientific advice, such as during the environmental crisis at Celco Arauco's Valdivia paper and pulp mill (chapter 4). But the executive government at the time dismissed CENMA, unable to see that it might have been better positioned to provide an "official" or "independent" assessment of the situation.[24]

Meanwhile, the University of Chile's response to this hostile environment has been to reduce expenditures, even if this requires evicting CENMA from the university-owned campus it occupies. Located in a park in the foothills of the Andes, this site housed the education and philosophy departments until the military government closed them. The modernist buildings lay empty through the dictatorship, until CENMA arrived (Jaksic 1989). Some scientists were at first reluctant to use the campus; it was rumored that the military government had operated a clandestine detention center on the site or evicted slums from there. Since the late 2000s, starved for resources, the University of Chile has planned to sell the property. At a time when rival universities have been investing in high-tech, scientific campuses on Santiago's outskirts, the University of Chile has not had the capacity—financial, political, or human—to even envision such a project.[25]

The Market for Environmental Science Is Triumphant

Although the engineers who first dreamed of CENMA told JICA that the Chilean state cannot by law have in-house laboratories, this is not strictly true. The state has several such laboratories, created between the 1930s and 1960s to provide technical advice to the government on fisheries management (IFOP), soil quality (Natural Resources Information Center), agriculture (Agricultural Development Institute), or forestry (INFOR). Like CENMA, though, in Chile's neoliberal democracy these labs have also lost their staff and funding (for a discussion of the IFOP case, see chapter 3).[26] Instead of cultivating strategic advisers or specialized, internal expertise, the state hires scientific advice contract by contract through public tenders, operated through the mercado público, in which universities, CENMA, in-house laboratories, and for-profit consultants compete. Many scientists I spoke with were highly critical of these tenders because, as in the example of air quality monitoring discussed earlier, they privilege cheapness over quality. One scientist described them as "the worst thing to happen to Chile's environment," and another as "the greatest imbecility the world has ever seen." He also denounced a global double standard: while in the United States they have the Environmental Protection Agency (EPA), "everything [international organizations] tell us Indians in South America is that the market is the only thing that works and to get rid of everything else … but look how they funnel money into the EPA and all its scientific capacity, all paid for by the state."[27]

Even when the Chilean state tries to support science in the public interest, it privileges a market logic. This occurred at the agency Corfo, when in 2010 it launched a new tender to fund "public goods science," defined as information that markets fail to produce. The tender reflected an economists' textbook approach to public policy that recommends state funding to solve "market failures," but ignored the problem Corfo had identified: that without base funding, public laboratories were falling to pieces and therefore unable to provide the information the state needed to make and enforce regulation.[28]

At the time CENMA was created, another environmental lab, EULA, was also created with foreign aid to expand the country's environmental science expertise. Like CENMA, EULA was set up as an autonomous unit within a university—this time the University of Concepción. But unlike CENMA, EULA has successfully funded itself through short-term contracts; about two-thirds of its budget has come from contract work, mostly for industry, and about 70 percent of it for EIAs (Bernasconi 2008). This model is widely celebrated by those who conflate "good" science with commercial success. In 2008, for instance, EULA was featured in a book on "successful experiences of research centers" in Latin America, led by prominent regional scholar of science Simon Schwartzman. The chapter on Chile highlights EULA for its combination of commercial success and academic excellence, which the author attributes to specific practices: to assert their autonomy while working with industry, EULA scientists work only on projects that have an interesting academic component, are strategic for the country (such as HidroAysén; see chapter 6), and belong to a company committed to confronting the environmental damage it causes (ibid.). The author also praises EULA's director of twenty years, Oscar Parra, for leading by example. Parra is a respected scientist who is also committed to public service. Among other things, for years he served on Conama's Advisory Council. In these ways, the Chile chapter concludes, EULA balances consulting and academic work as well as basic and applied science.

But not all scientists agree with this optimistic appraisal. Some criticized EULA, as they did CENMA, for not producing the kind of basic science they value. They described EULA as "too applied," "too close to industry," or "too political" because it had taken positions for or against specific projects under EIA evaluation. One consultant complained that CENMA was so technical it had become irrelevant, while EULA was too political to be trusted.[29] An activist with scientific training complained that EULA "is not accountable to anyone and does not produce official data."[30] Within EULA, a leading scientist preferred to highlight ongoing needs over the achievements. He wished state officials would recognize the value of scientific expertise for public affairs, specifically the value of environmental science, and was particularly frustrated that for almost two decades the state had refused to accredit EULA's PhD in environmental science, calling it "too interdisciplinary."[31] As Chile's only environmental science PhD program, this decision curtailed Chilean students' opportunities to train locally in environmental science. These statements reveal a shared frustration among scientists that the options available to them, be it CENMA, EULA, universities, or consultancies, are neither rendering environmental science a respectable scientific activity among lawmakers and the broader public nor providing scientists like themselves with adequate organizations within which to participate in the nation's environmental politics.

Conclusion

Despite efforts by Chile's post-1990 senators to distinguish consultants from university scientists, the consolidation of market mechanisms over funding for environmental science has blurred this boundary. Chile's two environmental science institutes, EULA and CENMA, have become more like consultants than universities as they compete for contracts in the market for scientific advice. JICA's vision for CENMA as a source of internal scientific advice for government failed because the political leaders of the new democracy were committed to the logics of the market and a small state. So too EULA's original mission to do academic environmental science has faltered because its operating budget relies on EIAs done for industry and the state has long refused to recognize the value of environmental science, withholding accreditation of EULA's only PhD program. Both institutes have struggled to sustain a reputation for scientific excellence that is academic and independent.

It is also useful to contrast the experiences of EULA and CENMA. Though both have struggled to balance their work for industry, government, and academia, CENMA has had a more difficult time at it, despite beginning with more funding and staff. A leading reason for their divergent fortunes lies in the constraints facing their home universities. Unlike the private nonprofit University of Concepción where EULA is located, the state-run University of Chile that houses CENMA has not had the resources to support the center, financially or institutionally. After it became clear that few people in Chile's government, Conama, or the University of Chile valued its founding mission, CENMA has had to reinvent itself multiple times over: as a center for research, government advice, or national reference, with limited success. EULA has been spared this kind of forced renewal thanks to steady financial support from the University of Concepción.

Interestingly, in all its efforts to fulfill a perceived need, CENMA never tried to be a public interest lab committed to protecting the environment. This kind of lab would have required a different funding scheme along with linkages to NGOs and regulatory agencies. To many Chileans, including CENMA's original champions, this lapse is common sense: Chile's subsidiary principle—whereby state action is restricted to that which is expressly allowed by law—precludes any public entity from proactively pursuing an agenda. Following the subsidiary principle, the state cannot have a privileged relationship with any one expert or center of expertise, especially groups that have environmental *protection* as their main goal.[32] Even among those most committed to CENMA, for it to be a lab working in the public interest was simply inconceivable. Better for CENMA to be once removed from governance—kept at a safe distance as a private foundation at the University of Chile. Yet this model failed over time. As CENMA invested in ISO certifications and turned to contract work, it became more like any other commercial laboratory without the special status a public institution might have.[33] CENMA's trajectory is thus symptomatic of a form of governance in which the state plays no strategic role in the production of knowledge; leaving this to market mechanisms, state officials abstain from negotiating the public credibility or usefulness of science. This approach to the management of knowledge production limits the powers and responsibilities of state agencies.

Things could have been different for Chile. This chapter examined scientific advice in transition: from dictatorship to democracy, and from a progressive era to a neoliberal one. The first transition involved obvious disruptions to institutions. Dictatorship brought repression along with strict adherence to military principles of order and compliance. Democracy then restored liberal principles of respect for human rights, citizenship, and democracy, within the constraints of Chile's 1980 constitution. The second transition, from progressive principles to neoliberalism, represents more continuity than change between the dictatorship and the post-1990 democracy. But for most of the century before the 1973 coup, Chile's political leaders interested in science saw it as a progressive force. Accordingly, they introduced policies to build the state's technical capacity through organizations like Corfo, government laboratories, technology transfer programs, and the expansion of higher education. Challenges from conservative forces often set progressives back, but only momentarily; after all, progressives had international organizations and foundations like Rockefeller on their side.

Though Chilean science was always a small, underfunded endeavor, including when compared to scientific activity in Argentina or Brazil, before 1973 most state policies toward science tried to expand its scope and influence.

Initially, it seemed the transition from dictatorship to democracy would revitalize science. Global policies like the EIA required it, international funders like JICA and the Italian government supported it, and Chilean scientists wanted to participate in governance. But neoliberal principles displaced these progressive ideals, permanently changing the social contract for science in Chile. This shift is evident in a number of decisions discussed in this chapter. First, Chile's new democratic leaders never restored the University of Chile's resources so it may be, as it once was, the nation's flagship university, with a strong public vocation. CENMA's impending eviction instead suggests that the University of Chile is heading toward obsolescence. Second, in all their discussions on the environmental framework law, Chile's senators never expressed any interest or concern about the quality or credibility of environmental scientific information.[34] Their indifference is best illustrated in the composition of Conama's Advisory Council and their rejection of a special fund for environmental science. Third, public tenders for procuring scientific advice have been configured to privilege cheapness over experience, quality, consistency, or durability. These have consolidated the market for science, enacting the ideal of a state that goes without privileged advisers, even as consultants—who have benefited most from knowledge markets—have become economically and politically powerful.

Though neoliberal principles have been triumphant in many ways, their advance has been neither seamless nor constant, as others have also emphasized (Hochstetler and Keck 2007; Mathews 2011; Plehwe 2009; Steinberg and VanDeveer 2012). In the case of Chile, in the transition from dictatorship to democracy, science gained an institutionalized role in policy through the environmental framework law and new laboratories even as it failed to gain any special authority in Chilean politics. Furthermore, the new management of science compounded historical challenges Chilean scientists faced, such as resource scarcity, local rivalries, and low capacity to self-organize. One result of these conditions is unstable scientific boundaries. The next four chapters analyze these contests over scientific authority in specific institutional contexts, beginning with two agencies that—unlike Conama—predated the neoliberal turn in Chilean government.

3   Salmon Aquaculture in Crisis

In April 2011, CIPMA, an environmental NGO, organized a conference titled Green Options for Sustainable Economic Growth.[1] Throughout the day, the audience heard from recognized entrepreneurs, activists, and politicians about how to produce more sustainable consumer products. Speakers stressed what they called "Chile's hard reality": in their view, Chile was a poor country that "wants to grow" but has few natural resources. The speakers reminded the audience that although Chile does not have gas, oil, or coal, it has ecological diversity. From north to south, speakers rolled off Chile's "natural riches": the Atacama, Huasco, Central Valley, coasts, forests of Valdivia, and lakes and fjords of the Chilean Lake District and Patagonia. To tap into this natural potential without endangering it, the country needed a state with a long-term vision committed to supporting policies and investments that could transform nature into commodities for export.

Two dissenting voices interjected into this otherwise-collegial discussion. Senator Guido Girardi criticized participants' faith in environmental protection via markets driven by the demands of "ethical consumers." Blinded by their allegiance to the market as well as their lack of intellectual or analytic capacities, Chilean elites, Girardi cried, have nothing to say about the ethical challenges caused by rapid technological change, from biotechnologies to climate change. On the other extreme, industrialist Wolf von Appen reassured the audience that nothing needed to change. Chile was doing well. He illustrated this by talking about salmon: salmon exports were quickly recovering after a devastating illness. Contrary to popular opinion, this epidemic was not evidence that things needed to change. Von Appen explained that the crisis was not the industry's fault. Rather, the problem had been sent to Chile from Europe in the form of tainted salmon eggs.

The question of what caused the collapse of Chile's farmed salmon industry vexed elites. If environmental factors were to blame, the crisis exposed the country's limits to growth. Nature could not be intensively transformed into export commodities without eventually destroying the natural conditions that made those transformations possible. If, on the other hand, the crisis resulted from bad husbandry practices and negligent farmers, then it exposed an irresponsible business class that could be regulated. But if foreign sabotage had caused the crisis, as von Appen suggested, this was a sign of success: Chile's production was destroyed because Chile had been conquering competitors' markets.

Chile's performance in the global salmon market had indeed been spectacular. From the first trials of industrial salmon cultivation in the 1970s, salmon production grew exponentially. In 1990, Chile exported 10,000 tons of salmon; in 2005, it exported almost 400,000 tons (Alvial et al. 2012). In just over a decade, Chile became the world's biggest salmon producer after Norway. This industry appeared out of nowhere—there are no wild salmon runs in Chile—through work, technological adaptation, and capital investments. Concentrated in lakes and fjords, the farmed salmon industry claimed to have transformed a damp environment into a desirable food source. To boot, it created around 45,000 jobs in the Chilean Lakes District and on the island of Chiloé, where people had few other employment options.

These gains were threatened by the infectious salmon anemia (ISA) epidemic that swept through farms between 2007 and 2010. ISA is an influenza-like virus common to farmed salmon (Godoy et al. 2008). Sometimes a salmon might not show any symptoms before suddenly dropping dead. At other times a sick fish might lose its appetite and appear lethargic. Usually the virus kills all the salmon in one pen. In Chile, the ISA virus caused an estimated 25,000 people to lose their jobs as salmon production fell from 386,000 tons in 2006 to 98,000 tons in 2010 (Asche et al. 2010; Alvial et al. 2012). Although the virus does not affect human health, it certainly hurt thousands of Chileans who lost their jobs and sent the industry into tailspin as stocks for sale vanished.

Following the crisis, it became a truism that "the industry grew more rapidly than government regulations could cope with" (Alvial et al. 2012, 25; Little et al. 2015). A postcrisis report prepared by consultants, state officials, and industry staff noted that "for years the industry's technical and commercial success was not accompanied by matching research, monitoring, and regulation to guard against foreseeable biological risks" (Alvial et al. 2012, 72). A safer industry, the report concluded, required local research and development, national analytic capacity, local monitoring, comprehensive regulations, and strong enforcement. Social scientists, some aquaculture scientists, and activists had been voicing similar concerns for a while (Buschmann et al. 2006; Barton and Floysand 2010; Bravo, Silva, and Lagos 2007; Leon-Muñoz et al. 2007; Schurman 2004). Moreover, the industry knew the risks: ISA outbreaks occurred in Norway in 1986 and the early 1990s, in Canada in the early 2000s, and on the Faroe Islands in 2003 (Asche et al. 2010).

Scholars have explained the Chilean state's reluctance or inability to regulate the salmon industry as a result of neoliberal principles, including a commitment to export-led development and limited regulatory oversight (Barton and Floysand 2010; Schurman 2004). This chapter focuses on a different dimension of governance being reshaped by neoliberal values: the state's analytic capacities, on which many blamed the industry's collapse. Compared to other industrial sectors, salmon farming was subject to potentially improved EIA regulations that might have fostered analytic capacities. These included standardized baselines and annual industry-wide monitoring, overseen by two agencies with more resources and authority than Conama had at the time.[2] In addition, the number of aquaculture scientists and consultants along with the funding available for aquaculture research were large compared to other fields of science in Chile. These factors made the salmon sector comparatively fruitful for critical communities. The requirement for industry-wide monitoring was particularly important for both environmental outcomes and critical communities because it creates constant demand for scientific data as well as analysis of the kind that can sustain long-term relationships between state officials and scientists. Yet this network seemingly did not live up to its potential. Instead, neoliberal values shaped a mode of knowledge production that sustained distrust and skepticism in government.

This chapter examines what it means for the state to "lack analytic capacity" by looking at how experts and officials tried to translate the ecological concept of "carrying capacity" into EIA regulation. Salmon farming is an excellent case with which to reflect on the promises of CIPMA's "green options" approach to environmental stewardship. Farm-raised salmon promised to alleviate pressure on the world's stressed wild fisheries, but aquaculture practices pose many environmental concerns given the industry's reliance on industrial agriculture, chemicals, and pharmaceuticals. As Girardi warned, salmon aquaculture grew exponentially thanks to rapid technological change, posing difficult ethical questions like what sustainability means in this case or how to respond to the suffering of salmon (Lien 2015; Gerhart 2017). Each time the state approves a farm's EIA, it should be balancing economic and environmental goals—a task that requires officials to exercise their analytic capacities. Otherwise, "the market pulls the carriage," as many CIPMA participants decried (and von Appen celebrated).

Despite its importance, the notion of the state's lack of analytic capacity remains ambiguous. It begs the question, If not on knowledge, then on what foundation does the state make its decisions? This chapter opens with the historical trajectory of Chile's salmon aquaculture industry and the significance of carrying capacity as an ecological and regulatory concept. The chapter then explores the relationships of distrust that characterize science, industry, and the state, fueled by market "incentives." Dogged by their distrust of scientists, state agencies turned to what historian Ted Porter (1995) called "mechanical objectivity" to build their confidence in the data they relied on to regulate the industry. These efforts failed for different reasons, even as the industry itself overhauled its production practices to reemerge an outwardly cleaner and more profitable business.

Salmon Aquaculture in Transition

Before the ISA virus crisis, salmon farming seemingly demonstrated Chile's economic "miracle." The industry generated spectacular growth but had uneven impacts on workers, who gained new opportunities but often became locked into low-paying jobs with few rights or protections (Schurman 2004; Gerhart 2017). From 1990 on, democratic presidents prioritized development as an "economic imperative": the economy had to grow, diversify, and export more goods. As a result, "the economic imperative gave the [salmon farming] sector its *raison d'etre*, and a high degree of flexibility in its operations" (Barton and Floysand 2010, 743).

Given the economic imperative, environmental regulation lagged. Before EIAs were required, a tight-knit group in the navy, National Fisheries Agency (Servicio Nacional de Pesca, or Sernapesca, which enforces aquaculture policy), and SalmonChile (the industry's association) supported the nascent industry by approving farm concessions quickly. By the early

2000s, environmental NGOs were pressuring the industry and state to tighten regulations (Barton and Floysand 2010). In response, Sernapesca and the Subsecretariat of Fisheries (Subsecretaria Nacional de Pesca y Acuicultura, or Subpesca, which proposes aquaculture policies) introduced disease control measures, protocols for hazardous waste disposal and pharmaceutical use, and the Environmental Regulation for Aquaculture Act (RAMA), which tailored EIAs to the needs of the aquaculture industry. Henceforth, prior to starting operations, farms would have to submit a declaration of environmental impacts with a standardized baseline report plus annual environmental monitoring reports. Between 2003 when RAMA entered into force and the ISA virus crisis, if a farm reported three consecutive negative monitoring reports, Subpesca could shut the farm down (Halwart, Soto, and Arthur 2007; Fuentes Olmos 2014).

The rapid pace at which new farms were approved, however, eclipsed the effects of these provisions. State officials had to meet monthly targets, much like US bank employees under pressure to grant household loans prior to the subprime mortgage crisis. Between 1997 and 2010, aquaculture farms accounted for almost one-quarter of all industrial projects that received EIA approval—just under four thousand farms. Many more farms were approved prior to 1997 with no environmental regulation at all. At the same time that RAMA was being implemented, the government approved a national aquaculture policy that prioritized industry growth and simplified the administrative requirements needed to open a farm. This national policy was a huge success: the plan aimed to double production in a decade but reached this goal within three years (Alvial et al. 2012). By contrast, around that time, governments in Norway, Canada, and Australia introduced different measures to limit the growth of their salmon industries (Lien 2015; Holm and Dalen 2003; Halwart, Soto, and Arthur 2007; Asche et al. 2010).

As marine cage salmon aquaculture has boomed globally, so have its environmental impacts (Halwart, Soto, and Arthur 2007). Salmon farming has four types of negative impacts: pressure on wild salmon species that are displaced by or interbreed with salmon that escape from farms; toxic and predatory effects on other species; water pollution resulting from the buildup of nutrients; and heavy metal pollution from antifouling agents used to clean the cages (Holm and Dalen 2003). The first concern is less relevant to Chile because salmon are not native to the Southern Hemisphere. Nonetheless, as ecologists have noted, escaped salmon have negatively

impacted other local species (Niklitschek et al. 2013). State officials and
NGOs have focused their concerns on the toxic and water pollution effects
felt by nearby species, including microbial and plant species.

A typical salmon farm consists of suspended cages tied to the seabed with
cables. Each cage contains thirty-two to eighty thousand fish, and one farm
can include anything from six to twenty cages (Halwart, Soto, and Arthur
2007; Lien 2015). Every day, the fish consume pellets and medicines fed to
them mechanically or by hand. Anything the fish do not eat falls to the sea-
floor, along with feces. Uneaten pellets, medicines, and feces together accumulate beneath the cages, adding nitrogen, phosphorous, salts, and chemicals
to ocean sediments and the water column (figure 3.1). This stock of organic
materials provides a home for bacterial communities that further deplete
oxygen in the water and release toxic gases, like hydrogen sulfide, nitrous
oxide, and methane. The accumulation of organic material and its associated

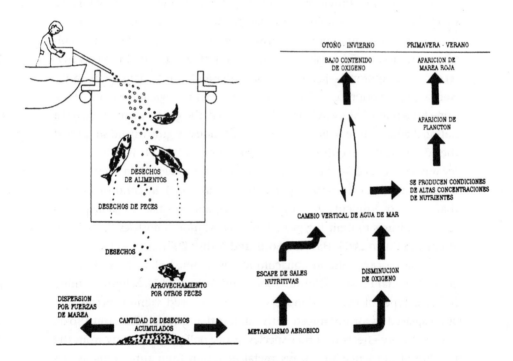

Figure 3.1
Illustration of the concept of carrying capacity at a salmon farm.
Originally printed in an industry magazine, *AquaNoticias Internacional*,
in September 1989. Reproduced with permission from Editec.

bacterial blooms have multiple negative environmental impacts. Plants and animals, including salmon, are deprived of the oxygen they need to be healthy. At the same time, algae and plankton thrive on the excess nutrients, crowding out other species and producing harmful blooms that further reduce dissolved oxygen levels. The toxic gas bubbles that float up to the water's surface transport bacteria that can infect fish, including salmon. Accordingly, Chilean scientists report reduced benthic biodiversity and an excess of nutrients around farms (Buschmann et al. 2006; Leon-Muñoz et al. 2007).

These processes are frequently summarized under the concept of carrying capacity. RAMA, the environmental code for aquaculture, requires that farms operate within the *capacity limits* of the body of water where the farm is located (article 74). Variably called carrying capacity, holding capacity, or water capacity, the concept assumes that the natural environment can only support a certain amount of production before water and sediment quality can no longer sustain healthy populations of microbes, plants, mollusks, and fish. Successful regulation thus requires "defining the capacities of bodies of water to support salmon farms ... [and] set limits on the maximum production in farming areas. Unless this is done, conditions will deteriorate leading to poor fish performance and eventually to disease" (Alvial et al. 2012, 4). But lawmakers have struggled to translate carrying capacity from a legal and ecological principle into actionable regulation.

The notion of carrying capacity has forced advocates of increased regulatory oversight in the farmed salmon industry to grapple with certain conceptual complexities. Carrying capacity can be viewed as a problem of equilibrium, population, or geography.[3] Though these are not mutually exclusive and may even reinforce each other, individuals I interviewed tended to privilege one conceptualization over the others. Chilean regulation currently privileges the first approach; RAMA requires farms to monitor pH, redox potential, and qualitative assessments of sediments, and take corrective measures if these fall out of balance and produce anaerobic (oxygen-free) conditions. By contrast, those critical of the state's regulatory approach emphasized the second conceptualization. They saw carrying capacity primarily as a problem of stocking densities. One scientist-turned-regulator described carrying capacity with a school yard metaphor: like children in a playground, fish are happier when they have room to run and are more likely to get sick if packed closely together.[4] He, and many others, was convinced a cleaner environment and healthier fish would result only from capping the number of fish per cage.

The third approach, also stressed by critics of the state's aquaculture regulations, focuses on geography: some sites are better for farming than others because of how quickly the water exchanges, and how organic materials are dispersed and diluted. Though this approach guides policy in Norway (Holm and Dalen 2003; Taranger et al. 2015), it was taboo in Chile because it promised to impose serious limits to growth.

Under pressure from the industry's collapse due to the ISA virus, state and industry officials reinforced RAMA's provisions (Fuentes Olmos 2014). Subpesca and Sernapesca also embarked on initiatives to improve their analytic and enforcement capacities. They sought to work more closely with scientists, changed the terms of their relationship with consultants, and studied how to better analyze environmental data from farms so as to strengthen the scientific foundations of their actions. In parallel, industry pursued measures that required farms in a given area to coordinate their husbandry practices. Whatever benefits these measures may have had for the environment, they were not based on a shared understanding of carrying capacity. Rather, the three visions of carrying capacity—as a problem of equilibrium, population, or geography—continued to coexist in Chilean aquaculture regulation. These conceptual differences persisted in part due to the continued lack of scientific knowledge about carrying capacity, both among university scientists (next section) and environmental consultants (the section that follows).

University Aquaculture Science

Given the importance of carrying capacity to aquaculture regulation, I assumed I could learn about it in books at university libraries. Several Chilean universities offered specialized degrees in aquaculture, and I targeted three of these for my search. To my surprise, no library contained more than a handful of Spanish-language volumes, and only one of these volumes contained data from Chilean farms—a book called *Aquaculture in Chile: 1983–2003*. While some books focused on other ecosystems (e.g., Spain) or were outdated (e.g., from 1982), *Aquaculture in Chile* contained recent data on production, genetics, fish health, and farming practices in Chile. Heartened by the book's seemingly rich empirical content, I continued reading in search of information about carrying capacity and how it applies to Chilean ecosystems. But I was disappointed; *Aquaculture in Chile* did not define carrying capacity or provide any information about it. What's more, it flatly denied that aquaculture has any environmental impacts at all.

I then noticed that the book listed no authors. Instead, it opened with a statement by President Lagos celebrating aquaculture's contributions to national development. The book was sponsored by pharmaceutical companies and published by TechnoPress, the publishing arm of Fundación Chile, a public-private organization heavily invested in salmon farming (Barton and Floysand 2010). *Aquaculture in Chile* is evidence of the kinds of analytic and intellectual "capacities" the state's commitment to economic growth over environmental concerns has engendered in Chile: limited and blind to the realities of the natural world. This book sponsored by industry and validated by governing elites was the only Spanish-language text available to students of aquaculture in Chile in their university libraries, highlighting the empirical lacunae students and their professors live with.

A scientist at a national aquaculture research institute associated with a prominent university confirmed that carrying capacity was not well understood in Chile in scientific terms:

> We [scientists in Chile] need to know what the carrying capacity of every fjord and estuary is. That is, how much production can that space tolerate without harming the environment in ways that negatively impact production. These studies don't exist and no one is doing them in Chile. Some studies that have tried to do this don't take into account all the relevant impacts.... For example, they don't analyze benefits from mussel aquaculture.... Without studies that take all these things into account, these studies are not useful for policy decisions. But that requires that people in government know what they want to know in the studies they ask for.

The ability of academic science to speak to policy was hamstrung, according to this biologist, by scientists and state officials who lack the training and vision to "know what to ask for." In addition, she complained, aquaculture science in Chile had been held back by industry's preference for veterinarians over biologists and oceanographers. Neither the state nor industry was interested in funding the studies in biology, oceanography, or ecology needed to understand carrying capacity.[5]

The gaps in scientific knowledge were seemingly not for lack of funding. Aquaculture science is relatively well funded in Chile. According to one assessment of aquaculture science, the state spent about US$153 million on aquaculture research between 1983 and 2005.[6] Of this, approximately US$30 million financed salmon-related projects and US$13 million went to environmental sciences. This adds up to almost 4 percent of the value of aquaculture-generated wealth between 1996 and 2004, compared to less than 0.6 percent spent nationally on all science as a proportion of national wealth (Bravo, Silva, and Lagos 2007, 258). The aquaculture scientists who wrote this national assessment further found that at least for one year in 2004–2005, Chilean funding for aquaculture science was on par with expenditures in Norway and the United Kingdom. The differences lay in how and for what Chile allocated funding.

The authors discovered that in Chile, funding focused on research into new species for farming to diversify the economy and develop new markets. By contrast, other governments spent research money on solving problems related to environmental impacts, disease control, and food safety (Bravo, Silva, and Lagos 2007, 122). Moreover, while in Norway, the United Kingdom, or Canada, a state-run committee of experts defines aquaculture research priorities, in Chile,

> there is no national scientific and technological policy that coordinates research and development funding in support of activities, objectives, and markets. More than a market failure, we see a system failure that can be resolved by creating an organization that directs, prioritizes, regulates, and evaluates research, to make public investments in science more efficient.... To date in Chile, there is no coordinating unit with access to all the scientific and technological information about aquaculture that could be used to identify and prioritize research areas in support of programs for a sustainable aquaculture industry, under the framework of the National Aquaculture Law. (Bravo, Silva, and Lagos 2007, 170)

Published just before the ISA virus crisis, this statement reflects the diagnosis repeated after it: the industry's success had not been "accompanied by matching research, monitoring and regulation to guard against foreseeable biological risks" (Alvial et al. 2012, 72). It also reflects CIPMA participants' hopes that the state might one day have the capacity to articulate a long-term vision of development based on sound scientific knowledge.

Why has it been so difficult to build the state's analytic and scientific capacities in these ways? It is primarily because of the widespread distrust of science and lack of interest from industry and the state. This is the conclusion of the national assessment of aquaculture science, based on focus groups with aquaculture participants from industry, government, and universities to explore what science does and should contribute to aquaculture. Speaking to me several years later, the subjects I interviewed corroborated many of the results from these earlier focus groups. These findings, together with the experience of two research institutes discussed next, suggest that historical legacies of underfunding, coupled with the conditions set by a market for scientific knowledge, fuel distrust in science among actors in government and industry.

Distrust in scientists runs deep among state and industry officials, who feel that scientists "opportunistically" accept research money to supplement their earnings (Bravo, Silva, and Lagos 2007, 186–187). A top-ranking state official formerly involved with Subpesca agreed wholeheartedly, explaining to me that "research funds are made to generate income for scientists at universities."[7] These misgivings reflect scientists' institutional constraints. Few scientists are employed full time so many cobble together an income from multiple sources, including consulting. Research money from Chile's national science agency is used to fund the research itself but also to supplement a scientist's otherwise-meager university salary. This kind of financial dependence exposed scientists to the suspicion that they acted opportunistically, accepting money to do research they perhaps did not have the equipment or expertise to do. Ironically, when discussed as policy, this same kind of financial dependence was part of an incentive structure oriented at encouraging scientists to do more work with industry (e.g., ibid., 123).[8]

State officials had an uneven relationship to science and scientists. Scientists often complained that state officials do not call on them for advice (Bravo, Silva, and Lagos 2007). This is borne out by the bibliographies of several reports prepared by Sernapesca (2005, 2008, 2010) and Subpesca (2002) during the 2000s; of a total of seven to twenty-four references cited in each report, between zero and two are by Chilean authors. The exception to this trend, a Subpesca (2008b) report analyzed below, documents monitoring data required by RAMA and other data collected by a navy-owned ship. State officials at both agencies, however, reported that they frequently called on Chilean scientists; staff members had no difficulty in identifying scientists they called on for advice.[9]

Industry also has been at best a reluctant client of aquaculture science. In the national assessment's focus groups, as in interviews, scientists complained that industry never asks anything of them.[10] Many felt personally slighted, and griped that scientists are hired through "friends and not your CV." Science was held back because scientists are not rewarded for their achievements but instead their social status—"we are still an underdeveloped country in this respect," this same scientist concluded.[11] Scientists also felt that industry, particularly multinational firms, disdain local efforts. Scientists imagined that staff members at foreign companies wonder, "Why should we trust the science these *negritos* [people of color] do, when in Norway we have the best science in the world."[12] STS scholar Myanna Lahsen (2004) found similar sentiments among Brazilian climate scientists, who felt they occupy the bottom of a global hierarchy, and express this through racial categories that ascribe whiteness to European scientists and color to South American ones.

Industry is especially uninterested in environmental research. The book *Aquaculture in Chile* and work of the Salmon Technology Institute (Intesal), the scientific arm of SalmonChile, both illustrate this fact. Intesal's work is centered entirely on fish health; a prominent consultant dismissed Intesal as an "institute for salmon health." He confirmed that when aquaculture companies are interested in science, they turn to veterinarians rather than oceanographers or biologists.[13] Industry participants seemed particularly uninterested in carrying capacity research. At one company, staff members explained that such research was slow and did not provide them with actionable information. Their company had recently declined to participate in studies to characterize the carrying capacity of one fjord where they operate because it was too expensive: they had to pay a fraction of US$500,000, shared between the government and several other companies operating in the same fjord.[14] Scientists largely attributed industry's lack of interest to its commitment to short-term profits, potentially threatened by ecological research that at best may generate benefits far in the future.

The state's relationship to science is mediated by government policies that promote science-industry cooperation at the expense of the public or state interest. That science needs to be economically productive was enshrined as policy in the mid-2000s, when the government introduced funding to support research in potentially high-growth sectors like aquaculture (Bustos 2015). Focus group participants likewise insisted that science be evaluated by patent counts and its impact on GDP growth—not other stated goals, such as improving the state's regulatory capacity or the industry's environmental performance (Bravo, Silva, and Lagos 2007). The policies that tried to push science closer to industry, however, have had mixed effects, as the next two examples show.

One of the beneficiaries of the new grant program was COPAS, an oceanography research center run by renowned scientists at a respected university. The grant required COPAS scientists to find matching funds from industry; being oceanographers, they looked to salmon farms. Building productive relationships with the industry proved difficult for a host of reasons, including distrust and the industry's lack of interest. But finding scientists interested in carrying capacity research also proved difficult; one scientist who finally got involved said, "I became an expert as I went along."[15] When I met her, she had recently finished research on one fjord in Southern Chile that is densely populated with salmon farms. Her research challenged several common assumptions about this fjord. She found its currents flow from coast to coast, not out to sea, meaning that water renews more slowly than previously assumed. The fjord was also a wind tunnel. Cages could be positioned to take advantage of these conditions to maximize the dispersal of organic materials. Nevertheless, her model did not produce a magic number—for instance, how many fish could safely live in one pen or how many farms the fjord could handle before suffering ecological harms. Industry and state officials therefore had yet to express interest in her results.

The second example involves IFOP, created in the 1960s with public funding to develop Chile's catch fisheries. Despite numerous organizational changes over the years, IFOP clung to its mission as a public research institute.[16] This commitment remained intact despite IFOP being reorganized in the early 2000s to produce "science for public goals." The reforms ostensibly aimed to support environmental science for regulators to use, but were accompanied by massive budget cuts that forced IFOP researchers to compete for research money. Because neither industry nor the state has much interest or trust in science, nearly all the competitive funding opportunities are run by Chile's national science agency, Conicyt. For IFOP scientists, this meant that as they competed against university scientists for research funding, their work became less relevant for regulation. Many lost their jobs, unable to compete against research scientists.[17]

IFOP's and COPAS's struggles for relevance illustrate how distrust in scientists grows among staff in industry and the state as all science, whether done for academic, industry, or public goals, is forced to compete for the same small pots of funding. Chilean scientists working at universities or research labs like IFOP or COPAS face incentives supposed to encourage them to undertake research outside their existing activities. But when they take these up, they are accused of being opportunistic for following the money. This paradox is compounded by the suspicions the market for science creates. For example, after the ISA virus crisis, when everyone became concerned that state agencies lacked the analytic capacity to regulate the industry, a legislator lamented the reforms introduced at IFOP, saying, "The State is in no condition to back-up regulatory requirements. Who will regulate? IFOP? An organization that is in practice broken, completely defunded, and that has to compete like any other university or institute for research funds to meet the State's needs?" (LH 20.434, 358). Like CENMA, had IFOP not been forced to compete for funding, it might have played the kind of analytic role many felt was necessary to avoid another ecological crisis.

Regulatory Science for the Aquaculture Industry

If not from universities or research institutes, then knowledge about carrying capacity might be found in state agencies or among environmental consultants. In 2006 and 2008, Subpesca (2006, 2008a) reported on underwater conditions using monitoring data collected by consultants on farms across the Chilean sea. As required by RAMA, the consultants collected data on indicators of anaerobic conditions such as pH, redox potential, organic matter, and others. Subpesca first presented this information in a short report that concluded that less than 5 percent of aquaculture farms (of any species) had anaerobic conditions. Two years later, Subpesca's second report again found that just 5 percent of farms had anaerobic conditions. This report also included over fifty maps showing concentrations of the indicators of anaerobic conditions on farm sites. The maps were based on data collected by consultants in 2005–2006, at the height of the industry's success, but were released in the midst of the ISA outbreak that killed millions of salmon at 238 farm sites—out of the 351 farms that were then operating (Mardones et al. 2011).

Anaerobic conditions indicate that the carrying capacity of a site has been exceeded, and the salmon and nearby fish are likely to be sick or weak from lack of oxygen, harmful bacteria, sea lice, noxious gases, and so on. Anaerobic conditions are also an indicator of high fish densities in the salmon pens—a known risk factor for ISA and poor salmon health in general (Godoy et al. 2008; Mardones et al. 2011).[18] The disconnect between the crisis unfolding before everyone's eyes and Subpesca's positive environmental assessment was bewildering, and raised doubts about the government's analytic capacity: How could environmental indicators be so rosy when the fish were suffering so much?

Subpesca's maps intensified doubts about the state's analytic capacities. First, the maps were difficult to read because the scale of the images obscured data patterns. Second, there were so many maps it was not clear which ones were important, whether they should be read together or not, or what they meant. Third, the work itself appeared sloppy. The data were more than two years old at the time of publication, and units sometimes changed inexplicably, for example, from "mg/L" to "ml/L," and back again. Fourth, the agency did not address issues that seemed crucial, like why nearly a quarter of the farms did not submit the required data or how many farms were at risk of returning three consecutive years of negative monitoring reports. This was, after all, the legal standard to shut a farm down.

To determine whether a farm had anaerobic conditions, Subpesca had to exercise analytic capacity in the form of judging when each indicator was too high or low, as the case may be, and then synthesizing these results across several indicators. Instead, Subpesca reported the data the consultants had submitted one indicator at a time and in seemingly arbitrary bands. Figure 3.2 is one example of over fifty similar maps Subpesca produced. The

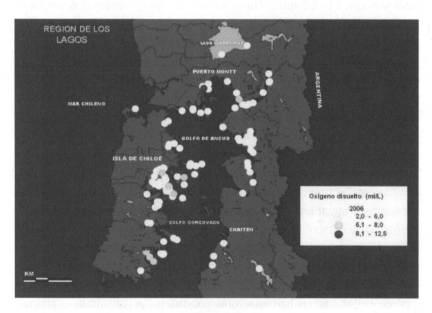

Figure 3.2
Dissolved oxygen levels at salmon farm sites across the Chilean Sea.
This map is one of 50 published by Subpesca in "Informe Ambiental de la
Acuicultura, 2005–2006." Although the objective of this map and others
like it was to facilitate identification of farms that had exceeded their carrying
capacity, the information is conveyed in a way that makes it difficult to do so.
Reproduced with permission from Subpesca.

Table 3.1
Cutoff Points for Indicators of Anaerobic Conditions

	Technical report 2002	Legal resolution no. 3612 of 2009
Organic material	≤ 15%	≤ 9%
Benthic macrofauna	Live organisms	n/a
Dissolved oxygen	≥ 0.5 ml/L	≥ 2.5 mg/L
pH	≥ 6.8	≥ 6.8
Redox potential (Eh)	≥ 0mV	≥ 50mV
Visual registry		No microorganisms or gas bubbles

map reports dissolved oxygen at sites around the Chilean Sea. The lowest range of dissolved oxygen is 2.0–6.0 ml/L, which is higher than the recommended 2002 limit (0.5 ml/L; see table 3.1). After the ISA virus crisis, the limit was raised to 2.5 ml/L. Thus, from a regulatory perspective, the band 2.0–6.0 obscures how many farms missed the limit or were close to missing it. This kind of obfuscation occurred with other maps too.

To dismiss these reports and the accompanying maps as merely sloppy misses how distrust shapes the state's analytic capacities. Despite the fact that Subpesca officials relied on consultants' data to prepare the maps, they actively distrusted both the consultants and data because of the consultants' financial dependence on salmon farmers. Aquaculture environmental consulting firms are specialized companies, typically organized around a recognized professional. The largest among them have their own labs and equipment, and hire several trained professionals. Relative to a salmon company, these consulting companies are small. Subpesca officials suspected consultants were at the service of the farms, so the agency did not publicize the cutoff limits it used to legally differentiate aerobic from anaerobic farms. The first column in table 3.1 reports only recommended cutoff limits, not the "official" standard Subpesca used to conclude in 2006 and 2008 that just under 5 percent of the farms had anaerobic conditions. A consultant explained that "the [cutoff limits] were an open secret. They weren't official. Because Subpesca managed the information with distrust. They would say to us, 'We don't trust you [consultants]. We know who is fine, who isn't, and so we control the farms. If we tell you the thresholds, you will manipulate the data.'"[19]

A Subpesca official concurred; because consultants work for salmon farms, telling the consultants what cutoff level differentiated healthy from unhealthy farms was like telling the industry how to score a goal against the state.[20] Subpesca staff members similarly explained that their 2006 and 2008 reports did not provide more detail, nor did they release each farm's raw data, because they suspected that consultants had manipulated the data. Subpesca and Sernapesca were so suspicious that consultants fabricated the baseline and monitoring data they even arranged for "an independent study" of underwater farm conditions that, they claimed, confirmed their fears: this independent report showed different conditions than those consultants had submitted. Like the cutoff limits and raw data itself, however, this study also remained secret. Instead of transparency, Subpesca used opacity to try to exert some control over a process of knowledge production the agency relied on but had no authority over (table 3.2).

Subpesca's strategy contravenes the assumption that numbers produce trust. The case of aquaculture farms is particularly interesting because each farm can be expected to perform to meet regulatory standards, thereby validating the standards while modifying farming operations to comply with those standards. This is an ideal scenario in a democracy given that the state's regulatory framework is validated at the same time that behaviors are directed toward the defined goal (Porter 1995). The numbers in this scenario objectify decision-making—a farm either has or does not have aerobic conditions—and facilitate accountability. The state can then easily impose sanctions on farms that miss the mark and thus deftly "administer the empire" (ibid., 224) by issuing regulatory standards that are easily validated. Studying cases like this across bureaucracies in the United Kingdom, France, and the United States, Porter argues that the language of numbers emerged in response to state agents' needs to communicate trustworthiness to actors outside their bureaucracies, therefore fostering democratic forms of accountability as numbers could be more easily used to hold agencies accountable for past decisions and actions. By contrast, in the Chilean case, rather than quell distrust, the language of numbers seemed to foster it. To state officials, the numbers only made lying easier for consultants and the industry they serve.

Subpesca then developed capacities designed to increase its control over consultants. The agency issued increasingly precise instructions for when, where, and how to extract samples from farms, so as to reduce consultants' abilities to take samples from sites that misrepresented farm conditions.

Table 3.2
Legally Required versus Publicly Available RAMA Indicators over Time

Legally required?	2003	2004	2005	2006	2007	2008	2009
Sediments							
pH-redox	Yes	Yes	Yes	Yes	Yes	Yes	Yes
Sulfur	No	No	No	No	No	No	Yes
Organic matter	Yes	Yes	Yes	Yes	Yes	Yes	Yes
Particle size	Yes	Yes	Yes	Yes	Yes	Yes	Yes
Benthic macrofauna	Yes	Yes	Yes	Yes	Yes	Yes	Yes
Visual registry (only hard seabed)	Yes	Yes	Yes	Yes	Yes	Yes	Yes
Water column							
Dissolved oxygen	Yes	Yes	Yes	Yes	Yes	Yes	Yes
Bathymetry	Yes	Yes	Yes	Yes	Yes	Yes	Yes
Currents	Yes	Yes	Yes	Yes	Yes	Yes	Yes
Publicly available?	**2003**	**2004**	**2005**	**2006**	**2007**	**2008**	**2009**
Sediments							
pH-redox	No	No	Yes	Yes	No	No	No
Sulfur	n/a	n/a	n/a	n/a	n/a	n/a	No
Organic matter	No	No	Yes	Yes	No	No	No
Particle size	No	No	No	No	No	No	No
Benthic macrofauna	No	No	No	No	No	No	No
Visual registry (only hard seabed)	No	No	No	No	No	No	No
Water column							
Dissolved oxygen	No	No	Yes	Yes	No	No	No
Bathymetry	No	No	No	No	No	No	No
Currents	No	No	No	No	No	No	No

Subpesca believed that if it could reduce consultants' discretion by forcing them to sample from predetermined sites, baseline and monitoring data would be more accurate. The agency thus introduced multiple requirements: it specified when during the year, how far outside a cage, and where exactly within the farm consultants should sample. Subpesca at one point asked that sampling sites be identified with GPS coordinates, presumably to verify the exact location from which each sample was taken. Consultants resented these measures because they signaled Subpesca's distrust of them, but also because they revealed how differently regulators and industry saw

farms, in a literal sense. During my interviews, several respondents spon-
taneously drew farm sites to illustrate their point. State officials invariably
drew farms as if one was underwater approaching the pen—a view that
emphasized their major concern: uneaten pellets and feces accumulating
on sediments below the cages (as in figure 3.1). Drawings by industry staff
members were similar, but added a tangle of lines to represent the many
cables that tie the cages to the seabed and protect them from predatory sea
lions as well as thieves.

By contrast, consultants' drawings portrayed salmon cages from the perspective of a boat, which is what consultants see as they approach a cage
to take measurements. From the water's surface, consultants see a tangle of
underwater cables and nets around the cages that threaten their safety. The
choice they face is how to obtain the best measurement without breaking
or losing expensive equipment. Given these realities, consultants claimed,
they cannot sample from immediately below the cages. Subpesca's efforts
to specify when and where to sample made no sense to them. Sampling
should not happen on a fixed calendar, as Subpesca required, but rather
right before harvest, when organic materials are highest. Instead of providing GPS coordinates, which Subpesca would never have the time or capacity to verify, consultants should be required to report how many feet away
from a cage the samples were taken.

In short, the distrust between Subpesca and consultants was mutual.
Consultants believed that Subpesca "gets stressed out with so much paper,"
and "takes these data reports and archives them. It has no time or people to
analyze them." In the consultants' view, it was the agency, not them, that
was lying, because the agency had claimed "we had a spectacularly great
aquaculture when the ISA virus was out of control."[21]

In regulating consultants' actions, Subpesca imposed what Porter has
called "mechanical objectivity," based on following the rules, as opposed
to "disciplinary objectivity," which requires consensus and specialization.
Mechanical objectivity is appealing because it can operate without resolving
conceptual disagreements, such as those over carrying capacity. Disciplinary
objectivity would instead require sufficient intellectual agreement to have a
shared interpretative framework linking farms, carrying capacity, and environmental impacts, sustained by a scientific community that can define,
measure, and represent concepts like carrying capacity. In the case of the
United States' Bureau of Land Management studied by Porter, by routinizing

a kind of mechanical objectivity, this agency increased its power and capacity over time. For Subpesca, conversely, mechanical objectivity produced a negative feedback loop: each attempt by Subpesca to regulate consultants' actions bred more distrust of the agency, which in turn became more anxious about its capacity to control the industry, engendering further efforts to regulate consultants, who then further resented the agency's efforts to control them. Subpesca could not routinize its preferred forms of objectivity over consultants' objections.

The following narrative from a consultant gives a sense of the depth of consultants' distrust of Subpesca and Sernapesca. Here the consultant sees the state as deeply incompetent, ignoring the experiences of consultants, advice of scientists, and published literature on the ISA virus:

> In 2006, before the ISA virus crisis, we [consulting firm] sent e-mails and photographs to Sernapesca showing sediment in poor conditions.... We never heard back. In March 2007, at a workshop, I spoke with Barry Hargrave, a big expert on these issues, and I asked him to ask Subpesca to raise the redox potential to 150, but I was told, "If we do that, all the salmon farmers will be on our case."
>
> Six months before the crisis exploded, Norwegian expert Are Nylund, hired by Mainstream [a large salmon company]... and with more than thirty years of experience on these issues, alerted the top salmon farmers. These top salmon farmers called themselves the G6, because they are so arrogant and hierarchical.... Nylund told me: the problem in Chile is that those in charge of the industry are clowns.... It was easy for the G6 to ignore Nylund; he is an old Norwegian, a crazy scientist, ironic, he mocks everyone. The G6 silenced him. And then, when ISA appeared, no one understood anything. No one had read even one paper about it.[22]

The ISA virus crisis shattered Subpesca's attempts to exert control over farms, consultants, and the environment. Among other measures discussed in detail in the next section, Subpesca tried to increase its trust in numbers postcrisis by regulating the market for consultants. This market was, after all, the major source of Subpesca's distrust of consultants and the data they collected. It opted to impose a public tender system, following a proposal made in 2005 by the United Nation's Food and Agricultural Organization. Under this system, the state hires consultants through a public competition, thereby severing the tie between consultants and salmon companies. As described by one official, "The new public tender system avoids certain distortions. Here and in China this is complicated. Our goal is to bring transparency to this process. Now, it will no longer be [the salmon farm] that hires its own laboratory, but we will do it, and this necessarily adds transparency."[23]

Consultants were outraged. They worried the new system would reward price over quality. They believed the first tender process—completed a few months before my fieldwork—confirmed their fears. The state capped bids at such a low price that the industry leader did not submit a proposal and the second largest firm undercut the smaller firms. Many believed the new system would force small consulting firms into bankruptcy, and "when instead of sixteen there are only three [consulting companies] left, then there will be collusion [between consultants and industry] and they will massage the prices as they like," cautioned one consultant.[24] Anger and fears aside, many consultants recognized problems did exist. One unhappy consultant also admitted, "I know there are [ISO] certified consultants who are not transparent. I also received pressures from clients of mine, asking me to present reports with false data. [I replied] 'I am so sorry but I can't do that; find someone else.' For this I've lost clients that others have won."[25] Several consultants shared similar stories. When asked how they prove to their clients or state agencies that they are credible, two (out of six) consultants I spoke to said they have proven their credibility by reaching results contrary to the interests of their clients; one proudly said their track record doing this was so strong, it allowed them to charge the highest prices.[26] And they also pushed back, complaining that "the authorities have no evidence. They have never checked on us, so who is at fault? Who can say my company has done bad work? Come and audit me."[27] The new tender system may have gained consultants credibility with state agencies, but this was not reciprocated.

Subpesca officials had other reasons to intervene in this market. Post-crisis, the agency had gained the legal authority to shut a farm down after only one bad monitoring report, not three as before.[28] This newfound capacity motivated state officials to regulate the market for consultants to cultivate their own, internal trust in numbers. An official explained, "Since [the standards were raised], to suspend production we had to be really sure of what we were saying. That is why we decided to give Sernapesca more control [over data production], and have cutoff thresholds that were more stringent and validated by everyone."[29] The pursuit of trust in numbers was thus necessary to assure a historically weak bureaucracy of its own capacity to act. It was not motivated, as in the United States (Porter 1995), by the need to convince rival state agencies of Subpesca's authority. Yet Subpesca might have had more success if it had accompanied its regulation of the environmental data market with public-facing acts of accountability geared at convincing a skeptical public of the agency's seriousness. This was a diverse public—it included Subpesca's network of knowledge makers (e.g., consultants and university scientists), a powerful industry hostile to environmental regulations, and a disillusioned public—unified only by its doubts in Subpesca's capacity to protect the marine environment from aquaculture. This episode, along with those described next, point to the hurdles Chilean agencies face in trying to position themselves as confident and capable.

Building the State's Analytic Capacity

Postcrisis, the state took steps to try to improve its own credibility among consultants, scientists, and industry professionals. Building trust in the state was desperately necessary in light of consultants' distrust. Consultants derided state officials who make laws from offices in Valparaiso, where the fisheries agencies are headquartered—a city with a perfectly shaped bay and wonderful weather. Officials, consultants scoffed, have no experience on the water, in beat-up boats, fighting the waves on Chile's stormy southern seas. One consultant laughed that "in Valparaiso, if they have seen a boat, it must be a cruise ship."[30] A former state official recognized that state agents discredit themselves because they write out requirements without knowing what they are talking about. Instead, "trust and quality are communicated by working in the field with industry, with [my] data, on [my] farms, and with [my] people. Quality is not communicated from Santiago or Valparaiso. From there, [state officials] lose credibility."[31] Thus, state agents were seen with incredulity by people in industry, consulting, and universities as well as members of the general public, with negative consequences for the state's ability to impose its decisions on others. Compliance is less likely and enforcement more difficult when confidence in the state's capacity is low to nonexistent, or conversely, the belief that state agents are incompetent is high.

One of the first trust-building measures Subpesca took was to organize a workshop to review the environmental indicators collected through RAMA with participants from universities, consultants, industry, and NGOs along with the state officials who organized it. Over two days, participants heard presentations about "the state of the art" regarding specific environmental variables in Chile's southern seas. They discussed the implications for RAMA, particularly how to improve this tool's effectiveness at regulating farms and preventing environmental harms. The workshop was a success in that participants agreed on several modifications that raised environmental quality standards; the cutoff limits that Subpesca had so cautiously guarded were finally made official by publishing them (table 3.1, second column).

But the workshop also revealed ongoing disagreements that illustrate the state's weak analytic capacity.[32] Participants disagreed on issues ranging from what needed safeguarding—just salmon or other species as well?—to the nature of carrying capacity. Some participants emphasized that the salmon themselves are the most vulnerable to the ecological harms RAMA seeks manage. Against this perspective, an NGO representative and some scientists spoke of the need to measure impacts over a larger area, which would require research into carrying capacity understood more broadly, starting with more exhaustive studies of tides and currents. Another (related) source of disagreement concerned the location of salmon farms. Several consultants and scientists said that data records show that some sites have low carrying capacity because waters there are shallow or exchange slowly. Nonetheless, Chile's state agencies remained unable or uninterested in tackling legacy farms that received concessions in ecologically inappropriate areas.[33]

Participants were also unsure about how to collect, synthesize, and analyze data. Several supported using a model or algorithm to integrate the disparate indicators of ecosystem health, but cautioned that they lacked "a scientific view" from which to do this. Without an empirical and theoretical understanding of the marine ecosystem, participants had no scientific basis on which to evaluate a farm's environmental performance (Subpesca 2008b, 25). Moreover, every indicator that would go into such a model was suspect. Participants questioned why they should count small bugs and bacteria in sediments below the farms if no one was sure how to use the information, few trusted it as accurate, and there was no community of trained professionals to analyze the thousands of samples produced. Another cautioned against regulating sulfur because "without knowing the situation in Chile, it is difficult to determine an acceptable cutoff" (ibid., 24). Consultants and scientists questioned whether it was worth measuring pH, organic material, or tides. Participants finally decided to eliminate the monitoring stations that had been required on all farms because no one was analyzing those data. Regarding dissolved oxygen—fundamental to salmon health—participants debated how to best determine a cutoff level absent a scientific model or whether to measure this as relative change (as they do in Norway) (ibid., 20).

Together these statements betray intense doubts about farm- and place-specific knowledge (in a geographic sense) that undermined the state's regulatory efforts. This was again illustrated during a discussion about tidal measurements. Participants could not agree what legal language to use to specify where these measurements should be taken. A state official asked whether equipment should be deployed at a "representative" point of the farm, as the existing resolution stated. Yet as one scientist in the room asked, What is representative? Is it the center of the farm or elsewhere? Perhaps, another state official suggested, it should say "the best area for measurements." This continued until another scientist interjected, "Often what we think [of a farm or ecosystem] is not representative of reality" (ibid., 15). Echoing Porter's notion of mechanical objectivity, he supported drafting the regulations in ways that did not leave room for interpretation.

Participants did not invoke a need for more funding to respond to these uncertainties but instead appealed to a need for coordination and analysis. One scientist encouraged "more dialogue among the industry's actors" because the information available at the time, produced through RAMA requirements, did not "allow us to be satisfied with current environmental conditions" (ibid., 12). His colleague concurred; information will always be missing but the important thing is to integrate what is known, so greater contact between the scientific and business communities was essential. Another scientist added that scientists need to have access to data held by the private sector, such as at SalmonChile's Intesal lab, but "we must also work together." Hargrave, the workshop's keynote speaker whose affiliation was listed only as "Canada" in the summary documents, also thought that "you all should analyze all the information and collect all the data to see different situations in different places, trying to make a map and look for the relationships between one area and another" (ibid., 4).

Subpesca and Sernapesca's other trust-building efforts picked up on these calls. Following the ISA virus crisis, Sernapesca's staff increased by a quarter, for a total of 493 employees, and its budget increased to over US$30 million. Subpesca's staff nearly doubled, and its budget also increased (Alvial et al. 2012). With these resources, Sernapesca acquired boats and equipment, making it possible for agency officials to visit farms on their own terms.

According to the same state official who explained why it was important to regulate the market for consultants, "In addition, now Sernapesca [staff] get on the boat [with consultants] and we see directly what [the consultants] are doing."[34]

After the crisis, Sernapesca replaced Subpesca as the agency responsible for analyzing baseline and monitoring data to determine which farms had anaerobic conditions. Many saw the ISA virus crisis as evidence that Subpesca could not be responsible for both approving farms—thereby promoting the industry's growth—and regulating their environmental performance. To prepare for this handover, Sernapesca published a report laying out its vision for how to handle monitoring data in light of the problems the ISA virus exposed. The report detailed a number of familiar problems: lax reporting and performance standards along with fragmented environmental and sanitary information that state officials could not integrate. Sernapesca also complained that "data collection [for monitoring] is a number on a paper or a register on an excel sheet that is submitted and processed by hand. Additionally, the number is processed late (possibly after months of delay), and there is no trust in the number (measurement) or its origin (geographic place)" (Sernapesca 2009, 19). The same was true for data on fish mortality. With such dispersed and untrustworthy data, state officials could not forecast environmental conditions on farms, identify emerging problems, or take preventive decisions.

Sernapesca proposed a system to potentially overcome these problems. The solution was to create a "(computer-based) information system that could capture, concentrate, and analyze the totality of data that because of different existing regulations, were now distributed across different departments and units" (ibid., 4). This includes not only environmental indicators required by RAMA but also health and hygiene indicators as well as other environmental factors, like algae blooms, required in other regulations. Funded from a program called "modernization of the state's administration," Sernapesca's proposal laid out an ambitious vision for a computer-based system that could integrate all this information:

> In conclusion, this report proposes a system that captures, concentrates, and processes all the information (environmental, health, and hygiene) relevant to salmon farming that flows through Sernapesca. This information would go into a model (mathematical or intelligent) that will have the capacity to diagnose in order to orient enforcement actions. In addition, this model should support

prevention and mitigation actions that [prevent] the dispersion of negative environmental and/or sanitary elements, that might cause salmon farms to lose production, leading to losses of employment and the global competitiveness of the industry. (Ibid., 51)

Sernapesca's proposal marks, on the one hand, a major shift toward transparency and accountability (which contrasts with Subpesca's earlier actions). The agency would occupy the center position of this so-called intelligent information system, and would mobilize these resources in defense of the environment and salmon health, which Sernapesca understood to be intimately connected. Yet on the other hand, Sernapesca's report goes into detail about how infeasible this intelligent system is. First, it is too expensive: monitoring equipment alone would cost over US$2 million. Second, according to Sernapesca's research of global practices, the plan seemed unnecessary: nothing like it exists in Norway, Scotland, the United States, or Canada. In these countries, Sernapesca found, the state does little to monitor the environmental impacts of salmon farming. Farms are instead required to monitor and report several aspects regarding health and hygiene, and above all the number and frequency of fish deaths along with the causes.

Sernapesca's report is all about trust: how to generate, maintain, and communicate it. It is a vision oriented toward allaying fears—held by staff members in industry, consultants, university scientists, activists, affected communities, and some state officials—that the state cannot be relied on to regulate a powerful industry. And it seeks to inspire trust through an imaginary of control and competence. Control is achieved in this vision by centralizing data and making one agency squarely responsible for the entire decision-making process. Competence is achieved by being up to date; hence the emphasis on real-time monitoring and online reporting, as well as appeals to a model. Not only is the model described as intelligent and mathematical (Sernapesca also refers to developing an algorithm to make decisions), but it also imposes a shared understanding of what variables matter most to salmon and environmental health.

Neither Sernapesca's intelligent information management system nor Subpesca's workshop resolved participants' distrust of the data used in environmental regulation and enforcement. Neither did they challenge the core premises of the state's regulation of salmon aquaculture. Participants and the report's unnamed authors were certainly critical of RAMA and the information collected through it, but their critiques led them to seek to strengthen RAMA by improving information management. Bigger questions, like what caused the spread of the ISA virus or how to best translate carrying capacity into regulation, were not answered; despite the crisis, the workshop, and new reports, participants did not work toward a shared vision of what had caused this epidemic. Thus industrialists like von Appel could continue to blame it on "foreign saboteurs." It was as if each group of professionals retained their own different perspectives of farms, as illustrated in their divergent drawings of salmon cages. By contrast, the industry overhauled its approach to carrying capacity by forcing farms in a given area to coordinate husbandry practices.

Coordinating Industry

To those in industry, the efforts detailed above mattered little when compared to the introduction of *barrios*, or neighborhoods, to coordinate fish husbandry practices across all the farms located within a specified zone. Industry staff and politicians alike championed barrios as an effective way to deal with carrying capacity. The idea likely came from SalmonChile and was legitimated by an ad hoc working group composed of high-ranking officials, a representative of Fundación Chile, and a scientist. Called the Salmon Roundtable, this political entity took SalmonChile's proposals for containing the ISA virus and future diseases and brokered its approval through Chile's legislature (Alvial et al. 2012). President Bachelet (2006–2010, 2014–2018) called barrios "one of the most recommended measures to respond to epidemics like the ISA virus, to avoid future events like this, and reduce their impacts if they do occur" (LH 20.434, 12). Senator Antonio Horvath said that with the barrios, farms would be so shipshape that entering a farm would be like "entering NASA" (LH 20.434, 330).

For farmers, barrios represented a major conceptual shift. As described by José, a low-level farmhand who lost his job in industry but found work with Sernapesca, before the ISA virus, "if someone from the outside came with some bug [into your farm] you were screwed.... The idea was to isolate yourself so no bug could reach you." Thus, before the ISA crisis, companies used to think, "I trust myself and no one else."[35] Barrios challenged this ethos by imposing coordinated husbandry practices within a given area; these practices include required fallow periods, planting and harvesting all fish at the same time, and stricter sanitary standards regarding how to dispose of wastes and dead fish. Legislators also required that farms now be spaced a mile and a half apart to make it harder for pathogens to spread. These practices are used in other salmon farming countries (Holm and Dalen 2003; Taranger et al. 2015), and many in Chile attribute the industry's recovery since 2009 to this suite of changes (Fuentes Olmos 2014) as well as the industry's move to cleaner waters in Chile's Patagonia region (Niklitschek et al. 2013; Estay and Chávez 2015).

Nevertheless, a question remained: Would barrios be effective if these practices were coordinated across farms that do not share the same water space? Water, and everything it carries, does not respect political boundaries. In the words of a scientist and former state official, "Barrios are fine to the extent that the zone that you finally create is an area where the measures you apply make sense. But many imagine zones that do not actually exist...because you are separating one area from another when they are essentially the same, so the measures should be the same."[36] To many in industry, however, barrios would still bring environmental improvements by introducing better husbandry practices. Benefits to the environment that might be foregone from creating "imaginary" barrios were not important to the farmers who had faith in the husbandry practices being introduced.

For small and midsize farmers, however, barrios imposed financial costs. If a producer's farms fell into one single area, they would face a long fallow period with no commercial activity. From the farm deck, José could see things clearly:

> The large companies are integrating vertically to isolate themselves even more. A mile and a half is not very much between farms. If your neighbor's operation belongs to a small company, it won't have the same conditions as yours, right? It won't operate the same way. Your value chain is integrated, you sell directly abroad with your own name, but he sells to third parties. These small companies were absorbed [by the large ones] after the ISA virus crisis.[37]

Another thing that stands out about barrios is that Subpesca and Sernapesca played almost no role in their creation. In sidelining these agencies, the new law illustrates the privatization of regulation, brokered through a partnership between industry and high-ranking politicians, that both reflected and reinforced fears that the state agencies were incompetent. Instead, industry delineated barrios as it saw fit, combining personal, business, and some technical criteria. For example, farmers who did not get along were not put in the same barrio, and farms belonging to the same company were divided between barrios to avoid cash flow problems. One consultant who worked on a proposed map was unapologetic about the absence of any scientific foundation to their work: these kinds of business considerations were more important to the success of barrios than currents and water flows.[38] Industry actors were accountable only to themselves in drawing barrios. The maps were not publicly available, the methods and proposals considered were not openly discussed, and the criteria used to allocate a farm to a barrio were never specified (Alvial et al. 2012).

Industry's approach to drawing barrio boundaries reflects its general attitude to the role of science in policy: it is best kept on the margins (Bustos 2015). The Salmon Roundtable included only one scientist—a well-known expert in algae who "knows nothing about salmon."[39] Critics were convinced he had been picked to participate precisely because he had the "wrong" expertise. Further evidence of the industry's indifference to science is found in the Salmon Roundtable's (2009) final report, which proposed to dedicate just US$50,000 to research, divided across twenty-four projects, with carrying capacity at the top of this list.

Conclusion

To many industry observers, the crisis precipitated by the ISA virus was only a matter of time. Intensive and novel farming impacts fish and their environment in unsurprising ways, but ones that are still difficult to predict (Soluri 2011). Nevertheless, in addition to the enormous loss of fish life and harm to workers' livelihoods, this story was remarkable to many people because of what these events revealed about the intersection between government and industry in a developing country seeking to dominate an emerging "green" market. Farmed salmon promised to satisfy affluent consumers' demand for healthier foods while saving the world's stressed wild fisheries. But at this scale of intensive farming, are harvested salmon still sustainable? Are farm-raised salmon a healthy food option when grown in oceans polluted by nutrients and heavy metals, in which native species find it increasingly difficult to survive? And at this scale, don't we humans have a responsibility to care for the fish, to minimize their suffering and maximize their welfare (Lien 2015)? The scale and success of Chile's salmon boom was enabled by technological progress in veterinary sciences, production

techniques, and transportation. As Senator Girardi warned CIPMA's participants as they explored the market's green options, this pace of technological change poses ethical questions that states and societies struggle with,
particularly when the information available for decision-makers is as shaky
as that described in this chapter.

The ISA virus crisis exposed relationships between state officials and
knowledge producers characterized by deep misgivings. Above all this case
shows that scientific knowledge produced through market transactions
inspires distrust among the consumers and producers of this knowledge.
The problem of distrust in market-generated data is thus not limited to
actors on the fringes of society, as activists or environmentalists might be.
Rather, it is a fundamental problem that affects state agents and knowledge
professionals, and that reinforces state agencies' weaknesses: state officials
have to make decisions using data they are skeptical of and are also limited
in how to change the conditions that underpin that distrust. This was true
of standardized baseline and monitoring data collected by specialized consultants as well as research produced by Chile's scientists. To state officials,
consultants depended on the salmon industry, and scientists—whether
they worked at a university, IFOP, or COPAS—opportunistically followed the
money; officials therefore found neither group was credible.

Subpesca and Sernapesca responded with different strategies to increase
their trust in aquaculture science, although none produced the desired
results. Some efforts aimed to increase mechanical objectivity in the data
by subjecting consultants to increasingly precise instructions for data collection and sampling. But consultants' opposition, grounded in their own
deeply held doubts about Subpesca's competency, thwarted Subpesca's
attempts to routinize these instructions. Other efforts, like the workshop
and the design for an intelligent computer-based system, targeted disciplinary forms of objectivity that require greater conceptual agreements and
shared norms of interaction among state officials and expert professionals.
The workshop in particular represents an effort to foster a critical community as scientists and consultants probed, alongside state officials, the
technical foundations of the nation's environmental regulations for aquaculture. Yet the event was less powerful than it might have been because
apprehension about local knowledge, due to perceived gaps, dubious
data, and the unavailability of ecological longitudinal data, undermined

scientists' and officials' willingness to articulate better standards. Repeated calls for more "coordination" or "integration" of "all the available knowledge" are best understood as wishes for a degree of disciplinary objectivity that has so far eluded regulatory science for aquaculture in Chile. Together these events suggest that when actors are apprehensive about the empirical material available to them, this will undermine state agencies' efforts to assert a reputation for sound analysis.

The new regulations that legislators approved after the crisis improved husbandry practices across the industry, but in ways that advanced the privatization of regulatory capacities. Industry led the proposal and implementation of the new regulations, which avoided issues many critics and regulators considered important, like stocking densities or currents and tides. The power of the Salmon Roundtable to propose and pass these measures also highlighted the weakness of Subpesca and Sernapesca, which fell short of an ideal form of governance that some interviewees preferred. When it came to aquaculture, the state still needed to, as one respondent put it, "draw the lines on the soccer pitch."[40] According to another scientist, the state's role is to write and enforce the rules: "When I move through the city, the state has to define where are the critical places to put stop lights. They don't go around asking me where I would like these to be put. They don't ask me because I am the user. It is the state's responsibility to implement regulation that will prevent crisis."[41] That fisheries agencies played a marginal role in making the new rules only aggravated a general skepticism of these agencies' capacities—a feeling many summarized with a peculiarly apt adage: "it is good fishing in troubled waters." The implication is that industry took advantage of the state's conceptual, empirical, and regulatory disarray, thereby heightening insecurity about what was known and not known about Chile's marine environment. These fears were confirmed in 2016, when the industry suffered an environmental and health crisis of similar proportions as well as causes to that of 2008 (Salmon Crisis 2016).

In this sector too a state official justified the state's limited capacities by alluding to the subsidiary principle: "the state can only do what the law allows it, while the private sector can do everything that is not prohibited to it."[42] In this light, the fact that Subpesca created a public tender system to regulate a market for science from consultants makes this case stand out compared with the next three. In these, although many people shared similar misgivings about a market for science, neither Conama nor its successor, the EIA Agency, ever attempted to intervene in the market for consultants. Outside of aquaculture, scientists and consultants who participate in EIAs are hired directly by the company whose project they study and assess. The next conflict, over the environmental effects of a new paper and pulp mill, pitted two rival teams of scientists, each working for a different patron and suspected of lacking autonomy to guarantee the quality of their work.

4 Proving Pollution at Celco Arauco's Valdivia Paper and Pulp Mill

In 2004, residents of the small city of Valdivia, in Southern Chile, fell under a cloud of putrid odors that caused nausea along with skin and eye irritations. They soon watched in horror as black-necked swans started falling from the sky into their backyards. The wetland sanctuary around Valdivia supported a population of several thousand black-necked swans; within six months, their numbers had dropped to three hundred (figure 4.1). Swans are normally large, proud animals, but the surviving swans appeared listless or nearly dead from starvation. The swans fled the wetland sanctuary as their food source, an aquatic plant called *luchecillo* (*Egeria densa*), disappeared. Those who died had waited too long to leave.

Local residents knew exactly who to blame: a paper and pulp mill that had begun operating just before the swans began to die. They immediately appealed to Conama, the environmental agency in charge of regulating the mill's operation, to shut it down. The scale and sadness of this disaster touched a raw nerve among residents; a local ornithologist called this "Chile's own 'silent spring.'"[1] To many locals, the timing of the mill's opening implicated it directly in this tragedy, as nothing else had changed in the wetland. For the next eight years, rival teams of scientists fought in court over alternative standards of evidence to prove or disprove that toxic pollution from the paper and pulp mill had destroyed the swans' food.

Environmental scientists are often called on to identify the causes of ecosystem change, particularly when that change is as toxic and sudden as it was at Valdivia's wetland sanctuary. Scientists have specialized knowledge and tools needed to detect toxins in the environment, and the ability to link these to environmental or human health effects. Scientists have been central participants in toxic disasters, from acute ones like the gas explosion at

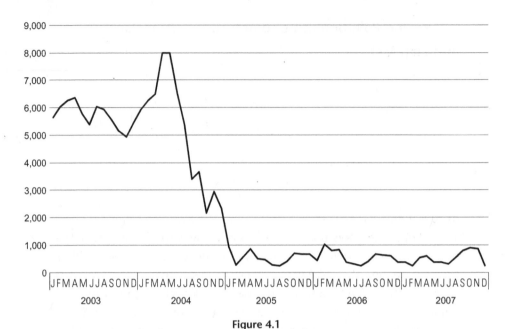

Figure 4.1

Census of black-necked swans in the Carlos Anwandter Nature Sanctuary, 2003–2007.
Data collected by Conaf (the National Forest Corporation) for public use.

a Union Carbide plant in Bhopal, India, to those caused by chronic exposure to chemicals like DDT or asbestos. They participate in multiple ways:
as advocates for victims or the environment, spokespersons for industry,
or seemingly neutral and rational observers of reality (Fischer 2000; Fortun
2001).

Reactions to Chile's "silent spring" were no different. In response to the
crisis, Conama and the company that owns the mill, Celco Arauco, each
hired scientists to determine if pollution from the mill had caused a toxic
effect on the swans or luchecillo. The scientists that Conama and Celco
Arauco hired differed in many ways, such as by their disciplinary training,
in the questions they were asked to research, and in the resources they had
access to. They differed also in their places of work and residence. Conama's
scientists worked at Austral University, located in Valdivia on the banks of
the Cruces River, which runs through the wetland sanctuary. Celco Arauco's
scientists instead worked at the Catholic University of Chile in the nation's
capital, Santiago. They also reached opposite conclusions. While the Austral University scientists found the paper and pulp mill responsible for

polluting the wetland sanctuary, the Catholic University scientists argued the proofs offered by their Austral colleagues were inconclusive.

One thing both teams of scientists had in common is that they were hired as scientific advisers through ad hoc, market-like channels. For that reason, each group was accused of producing "special interest science." As denounced by Rachel Carson and other activists, special interest science is that sponsored by industry to challenge public safety issues, from the effects of tobacco to pesticides and climate change (Dahl 2008; Oreskes and Conway 2011; Shrader-Frechette 2014). By definition, scientists who engage in special interest science are invested in that cause; their work is partial rather than neutral or objective. In Chile, participants in the Valdivia mill conflict expressed this concern with the adage, as mentioned earlier, "he who pays the piper calls the tune," used to denote the fragility of scientific autonomy where state agencies and corporations can hire scientists in any way they like. Because of these conditions, the controversy over what role, if any, the Valdivia mill played in the ecological collapse observed at the Cruces River expanded from the question of Celco Arauco's responsibility to encompass the credibility of Chilean scientists to speak to issues of public concern when they depend financially on one of the interested parties.

The Valdivia mill was the first large project to be evaluated by Conama through the EIA process, which in 1997 became obligatory for all industrial projects. The mill's EIA and ensuing ecological crisis therefore tested, for the first time, the credibility of EIAs and Conama, much like the ISA virus crisis tested environmental regulations for aquaculture along with the capacities of Subpesca and Sernapesca (see chapter 3). When it came to enforcement, the three agencies had some qualities in common: limited to nonexistent capacities for surprise inspections and no in-house scientific capacities. As with the fisheries agencies, Conama had to rely on external experts it hired through market channels. In this case, however, Conama and Celco Arauco relied to a much greater extent on university scientists than on consultants. As in the cases of CENMA and EULA analyzed in chapter 2, events and actions in this instance blurred the boundaries between scientific and consulting work, in the process obscuring potential differences between "special" and "public" interest science.

The chapter starts by examining how this controversy threw into doubt the ability of Conama to evaluate industrial projects and manage a contentious crisis. This involved one of Chile's most powerful business groups: the

Angelini family conglomerate, with business interests in energy and media. Anacleto Angelini, a self-made Italian immigrant, purchased Celco Arauco in 1977 from the state. He built a forestry empire that by the early 2000s included one and a half million hectares of plantations, six paper and pulp mills, eight wood panel mills, and several industrial sawmills in Chile, Argentina, Brazil, and Uruguay. Celco Arauco and the Angelini group are emblematic of the tight-knit relations that characterize Chile's economic and political elites (Escaida et al. 2014; Klubock 2014).

The chapter then explores the role scientists played in the political and legal battles to determine if the mill was responsible. When ecological crisis strikes, are environmental scientists credible participants, and if not, who is? How did participating scientists affirm their credibility in light of suspicions that they lacked autonomy and impartiality? What consequences did Chile's knowledge market have for scientific credibility? The analysis demonstrates how this controversy threw into doubt the credibility of the scientists who participated; their personal reputations suffered, as did those of their respective universities. Moreover, Chilean scientists, both those local to Valdivia and those from further away, depicted "good science" in increasingly narrow ways and made good funding synonymous with good science. Many scientists used a surprising argument to communicate their impartiality to others: they claimed their independence was evident because they reached results contrary to their funders' interests.

From the crisis, Celco Arauco emerged a new company. It created a new environment unit, implemented new models of scientific cooperation, and embraced the guilty verdict Chile's civil court handed it in a remarkable legal decision. In July 2013, Valdivia's first civil court found Celco Arauco guilty. The narrative closes by looking at how, rather than appeal, Celco Arauco accepted the verdict and set to work on a wetland remediation plan, but this time in partnership with local environmental NGOs and Austral University.

Conama Evaluates the Valdivia Mill's EIA

Worldwide, paper and pulp mills are a major source of industrial pollution (Sumathi and Hung 2005). Wastewater effluent from the pulping and bleaching process contains acids, suspended solids, dissolved oxygen, and organic materials that when discharged into waterways, produce environmental harms. Evidence of environmental harms from paper and pulp mills

Table 4.1
Timeline of Events

1996–1998		Mehuín residents oppose Celco Arauco's plans to dump untreated wastes into the ocean
1998–1999		Conama approves the mill's EIA with tertiary waste treatment and dumping into the Cruces River
2004	February	Operations begin; the mill is fined and briefly closed due to air pollution
	July	Swans are declining, says International Union for the Conservation of Nature
	August	Consultants report nineteen violations of the mill's permits, including unauthorized excess production and waste disposal capacities
	November	Swans drop from the sky and flee the wetland; social alarm is rising
	December	Austral report is released, accusing the Valdivia mill of toxic pollution, and the authorities impose more fines
2005	January	Conama temporarily closes the mill due to irregularities
	February	Conama reopens the mill; conflicting scientific reports are released by Zaror, Ramsar, Austral, and others
	April	Scientific reports continue to proliferate
		Two court cases are filed against Celco Arauco—one by the state prosecutor and another by citizens
	May	Conama modifies the mill's environmental permit, allowing it to continue operating
	June	The Supreme Court finds Celco Arauco not guilty in the case brought by citizens
2006	August	Conama authorizes the mill to increase production and the remediation plan stalls
2009	July	A judge confirms the fines Arauco received in 2004–2005
2010	February	Conama approves an EIA for a waste pipeline through Mehuín; protests there continue and (as of this writing) the pipeline has not been built
2013	July	Valdivia's First Civil Court finds Celco Arauco responsible for the events that led to the environmental collapse of 2004

was first documented in the 1950s. In the 1980s, pollution fell dramatically after alternatives to elemental chlorine were developed, and governments limited wastewater discharges and required mills to make use of the best-available technologies. These measures were supported by scientific studies conducted mostly in Canada, the United States, and Scandinavia that showed toxic reactions in algae, fish, and benthic species (Halliburton and Maddison 2004). Reports from the International Conference on the Fate and Effects of Paper and Pulp Mill Effluents, held every few years, explain that pulp mill toxicity is a function of the quality of the wood, how well a mill is run, and the pulping and bleaching methods used. A toxic reaction depends on the characteristics of the receiving body of water—for example, estuaries are more vulnerable than open rivers. As with other environmental issues, this body of scientific research shows that pulp mill toxicity results from both universal aspects, like pulping and bleaching methods, and idiosyncratic ones that are specific to a time and place.

Conama was responsible for evaluating these and other aspects of the Valdivia mill through the EIA. Conama was under pressure to approve the mill because it promised to contribute to economic growth (Camus and Hajek 1998; Sepúlveda and Villarroel 2012).[2] This case shows a new agency, Conama, making the best of the limited capacities it had. Though it would later be sharply criticized, the agency exceeded minimum requirements in some ways. For instance, it requested more information from Celco Arauco and, remarkably, that the company consider alternative means of waste disposal. Outside such requests, Conama's capacities were quite limited. By law, Conama could not object to the mill's proposed location or recommend alternatives.[3] When the Valdivia mill's EIA came in, the new agency had no staff with training in chemistry or experience evaluating such a large, complex industrial facility.[4] Conama's inexperience was so deep that the official who first received the mill's EIA later confessed that "he did not know how to proceed" (Sepúlveda 2016, 159). When the agency's evaluative capacities are exceeded like this, the regional office has to rely on external partners, such as assistance from the central office, in Santiago, and consultancies.

The central problem was always where to dump the mill's wastewater effluent. Celco Arauco first proposed to discharge wastewater directly into the Pacific Ocean through a waste pipeline that would have run into the Bay of Maiquillahue through the coastal town of Mehuín, a Mapuche indigenous community (figure 4.2). The residents of Mehuín, however, refused

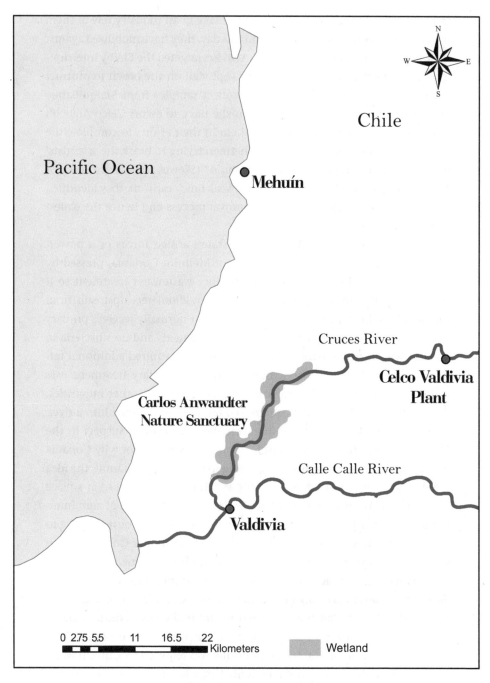

Figure 4.2
Map of the Valdivia area.

to risk the ocean and sandy beach for the sake of an industry few of them would ever work for; since 1996 and to this day, they have mobilized against the waste pipeline (Nahuelpan 2016). Activists targeted the EIA by interrupting data collection. For months, locals kept vigil on the beach to obstruct Celco Arauco's scientists' attempts to collect samples from Maiquillahue for the EIA studies. The state even sent the navy to escort Celco Arauco's scientists into the bay so as to assist them in their efforts to complete the EIA. The navy opened fire on local fishermen trying to block the scientists' passage, thus starting the "Battle for Mehuín" (Skewes and Guerra 2004). A Mehuín activist called their strategy the "weak link": early on, they identified the EIA as a weak link in the mill's approval process and hence the easiest aspect to disrupt.[5]

Ultimately, local residents—not the state's armed forces or a powerful national company—won the Battle for Mehuín. Conama, pressed by the protests, asked Celco Arauco to introduce wastewater treatment so it could dump this into the Cruces River, thirty kilometers upstream from the city of Valdivia. Paper and pulp mill effluent normally receives primary and secondary treatment that removes large elements and uses bacteria to break down organic wastes. In this case, Conama required additional tertiary waste treatment. Rare in the pulp industry, tertiary treatment uses aluminum sulfate to purify water by clumping together other impurities. The resulting solids can then be removed rather than dumped into a river, lake, or ocean. Although aluminum later became a prime suspect in the disappearance of the luchecillo, little is known about how or why Conama required this tertiary treatment. According to the mill's EIA permit, the idea came from a consultant identified only as "Homsi" who worked at a place called "KRYSTAL" (Conama 1998, 4). Despite the importance of aluminum in the case against Celco Arauco, Homsi never testified in court or spoke to the media, no one I spoke to knew anything about him or Conama's decision to follow his advice, and my efforts to find Homsi came to nothing.

Conama argued that tertiary treatment would minimize toxicity, to be measured by adsorbable organic halides (AOX), an indicator of toxicity in pulp mill effluents that was regulated in the mill's EIA permit. Conama hoped to dispel citizen's fears about pollution with these measures. Neither tertiary treatment nor AOX monitoring was required by Chilean water quality standards, so their use represented an effort to exceed the requirements. Generally, when it came to pollution control, Conama's capacities were limited. For example, the agency could not require the use of best-available technologies, and the EIA permits did not expire. Although the Valdivia mill's operations began six years after Conama issued its EIA permit, Conama could neither bring the permit up to date in terms of anti-pollution technologies nor request that the company renew the baseline studies. In this light, for Conama to introduce tertiary treatment and AOX monitoring was significant.

There were, however, a number of problems with Conama's reliance on AOX. Conama's own justifications for the measures in the mill's permit were vague. In response to citizens' concerns, the agency stated simply that tertiary treatment is "expensive" and AOX monitoring will be subject to "tight controls." The agency did not address scientific studies available at the time that demonstrated that zero-AOX effluent could still be toxic. Globally, scientists and policy makers were then in an open debate about how effective zero-AOX policies were (Halliburton and Maddison 2004; Lehtinen 2004). Worse, at the time no laboratories in Chile could reliably measure AOX. The mill's permit foresaw sending samples abroad for analysis—making it impossible to detect toxicity quickly or subject AOX to tight controls (Conama 1998, 17, 28, 71).[6]

Conama also imposed a demanding monitoring schedule. Monitoring is vitally important for detecting toxicity in light of idiosyncratic factors that affect paper and pulp mills (Dubé 2004). Conama (1998, 65–68) initially required that Celco Arauco monitor certain elements continuously, weekly, monthly, and annually, and subject the samples to several analytic tests. For example, each year Arauco had to analyze and evaluate the bioaccumulation of certain metals and toxic elements in several species, including the luchecillo, and monitor impacts on endangered species. Unfortunately, one year after issuing the mill's EIA permit, Conama (1999) reduced many of these requirements, including the monitoring of the luchecillo, as well as the obligation to install additional water monitoring stations at the wetland's entrance.

The above suggests that Conama tried to use the EIA process to introduce high operating standards for the Valdivia mill. But the agency then backtracked, and also made confusing statements that raised doubts about Conama's true intentions or grasp of the issues. First, the technical foundations of the permit appeared weak because the agency did not justify its decisions in technical terms. Conama did not explain, for instance, why it accepted the advice of the mysterious consultant Homsi, why it chose AOX (which could not be measured by local labs), or its selection of which species to monitor. Sometimes the lack of a technical justification would come back to haunt Conama; for instance, years later some scientists criticized the agency for not requiring Celco Arauco to monitor indicator or sentinel species that could serve as an early warning system of accumulating toxins or heavy metals.

Second, the permit was contradictory, and included statements that denied or minimized problems with toxicity. The permit states point-blank that "the effluent is not toxic" and oceans also emit organochlorides, implying that because these are naturally occurring substances, high concentrations from industrial operations are unimportant (Conama 1998, 13, 21, 24). The permit even cites an experience where fish lived in a river of 100 percent toxic effluent and notes, "Nevertheless, after more than fifty years of effluent discharge from almost a thousand paper and pulp mills in the world, into the rivers of Canada, the United States, and Scandinavia, important accumulations of organochlorides generated by the pulp industry have not been found in the ecosystem" (ibid., 18).

Globally, paper and pulp mills are known to emit toxic wastes that are at best difficult to manage (Halliburton and Maddison 2004; Sumathi and Hung 2005). Faced with this challenge, Conama's actions, which introduced crucial modifications and monitoring requirements based on studies the agency requested while simultaneously minimizing or denying potential harms, helped to erode trust in institutions. For example, the agency could neither manage nor correct misleading statements that political and business leaders made about the mill, such as assurances that the mill's "AOX levels would be normal" or that it would not use chlorine—a lie that for years confused members of the public (Escaida et al. 2014, 76–82). Statements like these caused public alarm and distrust in the government as well as Conama (Sepúlveda 2016; Skewes and Guerra 2004).[7] Scientists too became skeptical of Conama's abilities to manage industrial pollution given the contradictions in the EIA permit.[8]

Making Local Knowledge Uncertain

As the swans in the Cruces River began to disappear, Conama solicited several scientific reports. The most important of these was done by an interdisciplinary team of scientists at Austral University, located in Valdivia on the banks of the protected wetland. In five months, Austral scientists produced a report that answered the questions Conama posed to them: What had caused the swans' death and migration out of the wetland? Was pollution involved, and where from? How could the wetland's ecological state best be described, and would the area's environmental problems get worse? In many ways, Austral was the natural choice for this task: since its creation in 1954, Austral has tried to live up to its institutional motto "Knowledge and Nature." University scientists brokered the international protection of the wetland, and many see themselves as local actors committed to studying every aspect of Valdivia's natural resource-based economy (Escaida et al. 2014). Indeed, when crisis struck, the university was already so deeply implicated in the project that many residents of Valdivia worried Austral's ability to objectively evaluate potential wrongdoings was compromised (Sepúlveda 2016). Several Austral scientists had participated in the mill's original baseline studies in 1995–1996, some had raised concerns about the project in 1996–1998, and in 2004, a few of them sounded the alarm when the swans began to die.

Austral's inquiries initially focused on the swans, the most visible symbol of the wetland's environmental collapse. A group of almost thirty scientists trained in veterinary sciences, zoology, biology, ecology, and other fields set out to determine what caused the environmental collapse they and their neighbors had observed. In 539 pages that relied extensively on existing monitoring data, the Austral (2005) report concluded that the Valdivia mill was the most likely culprit: the scientists found high concentrations of metals like iron and aluminum in the water along with increases in pollutants, like dissolved inorganic wastes, and indicators of pollution, such as conductivity and dissolved oxygen. They concluded that pollution from other sources like agriculture or a nearby sewage treatment plant were too small to be toxic, and had not changed significantly. Neither were there any fluctuations in precipitation or river flow. Instead, the Austral scientists hypothesized that toxicity from the mill's effluent killed the luchecillo, leading the swans to starve and suffer iron poisoning. A gap in the Austral scientists' argument remained, however, because they could not specify exactly what—aluminum, chlorates, or turbidity?—had produced the toxic shock.

The Austral scientists became national celebrities (Oyarzo 2014). Their spokesperson was the team's lead investigator, zoologist Eduardo Jaramillo. Jaramillo received his PhD from the University of New Hampshire, and did postgraduate work in South Africa and Germany. But for most of the past three decades, he carried out his work where he lived: on the banks of the Cruces River wetland and nearby beaches. This proximity to the place allowed him to observe the swans' suffering from the get-go (Escaida et al. 2014, 104). Jaramillo would spend his weekends on the Cruces River on his motorboat. Even when sitting in his small office at the university, Jaramillo was outdoorsy, dressed in heavy boots, well-worn pants, and a wool sweater.

Jaramillo was no stranger to Celco Arauco or controversy. During the Valdivia mill's EIA evaluation, he was one of the scientists trying to get access to Mehuín's Bay of Maiquillahue, where he directed Austral University's coastal fisheries research station. Protesters forced Jaramillo and his whole research station out of town. One local in Mehuín told me that between 1996 and 1998, "there was a price on the head of Jaramillo."[9] Almost a decade later, some team members resented what they regarded as his heavy-handed style. But his biggest critics were rival scientists from other universities who condemned Austral's report and Jaramillo in particular. They claimed that Austral scientists had succumbed to public pressure; according to one observer, "[The Austral scientists] all wanted to be heroes," and Jaramillo more than any other.[10] Nonetheless, both in public and private, Jaramillo always spoke calmly and confidently about his team's conclusions.

Before Austral's complete report had been released to the public, Conama contracted additional scientific reports. One was by Claudio Zaror (2005), a chemical engineer who is a professor at the University of Concepción. Another was by the Ramsar Secretariat for Wetland Protection, which sent two foreign scientists that activists considered more credible because they lacked any relationship to local organizations. Celco Arauco meanwhile commissioned its own report from ecologists at the Catholic University in order to contest the results of Austral's work (CASEB 2005). A public relations expert explained that at the time, "[Chilean society] fell into a scientific fantasy."[11] Everyone wanted to believe scientists could determine, beyond a shadow of a doubt, responsibility for the wetland's collapse. As more and more scientists got involved, and in a context where the credibility of the mill's EIA permit and Conama itself were already suspect, how to interpret Austral's report became contested.[12] These challenges can be grouped in two broad strategies: accusations that the Austral report's

Table 4.2

Potential Causes of Toxicity in the Valdivia Mill's Effluent Identified by Different Actors in the Early Days of the Crisis

Actor	Potential causes of toxicity
Austral scientists	Aluminum sulphate
Ramsar scientists	Chlorine-based compounds
Claudio Zaror	Conductivity, sodium, sulphates, AOX, suspended solids, nitrogen, phosphorus, and resins
NGOs	Low water relative to effluent, chlorine, wood-based compounds, and dioxins

scientific proofs were inconclusive, and complaints that it was impossible to scientifically know anything with any certainty at all.

The industry-sponsored Catholic University ecologists focused their energies on punching holes in the Austral report; as if they were lawyers, Catholic University scientists sought to raise a "reasonable doubt" about the mill's responsibility (CASEB 2005). The university's short report criticizes Austral's science, particularly its sampling methods, use of certain analytic tools, "sloppiness," and failure to consider alternative explanations. Alternative explanations soon started to proliferate. Each report that came out in the early days of the crisis blamed a different element (table 4.2). Even activists represented by different NGOs proposed new explanations. (The Catholic University report is not included in table 4.2 because it did not try to explain events but rather only raise doubts about Austral's report.)

Within a few years, ten different hypotheses existed to explain the wetland's environmental tragedy: three chemical hypotheses; three physical hypotheses, based on stream flow and ultraviolet radiation; two biological hypotheses, based on ecological succession; and two "multivariable" biological hypotheses (*CDE v. Arauco* 2013, 3693).[13] In court, scientists testified that ocean tides, rainfall, sediments, and dilution were all important. Some argued that eutrophication in the river was low, and others asserted that it was advanced. In short, the most common way of contesting the Austral scientists' conclusion was to produce a seemingly endless number of untested hypotheses—a veritable "war of hypotheses" broadcast to the nation through news reports, TV interviews, and letters to the editor of major newspapers (Sepúlveda 2016, 252).

As with salmon farming (chapter 3), in this case too apprehension about local, placed-based knowledge fueled distrust in science. What stands out in this case is how publicly and enthusiastically many scientists participated in raising doubts about what was and was not known (Escaida et al. 2014). For example, the Ramsar scientists concluded it was impossible to know anything with certainty because, among other data gaps, estimates of water depth at strategic points were impossible to estimate with the available data. They denounced a collective "dancing with numbers" (Ramsar 2005; WWF 2005). Despite the swans' charisma, prior to the crisis, only one scientist had dedicated himself to studying the swans and luchecillo. His research, however, was often misrepresented by scientists eager to cast doubt on Austral's science and Celco Arauco's (alleged) responsibility (Sepúlveda 2016).

For Austral scientists, their doubts also stemmed from their misgivings about the quality of monitoring data collected by Celco Arauco. Their report suggests that the company lied about its monitoring results: the Valdivia mill's pollution reports to Conama did not match the laboratory reports for a range of elements, including pH, suspended organic solids, arsenic, copper, cobalt, and chlorophenols. Conama staff members were sufficiently aware of these discrepancies that they too distrusted the data they had.[14] These discrepancies were confirmed by Zaror, hired by Conama to review the mill's operations. Zaror found that for two months, the equipment to measure conductivity—an important indicator of pollution—was broken. Furthermore, Celco Arauco staff members had been underreporting chlorophenols discharges because they used the wrong methods to measure resins and acids. And he discovered that water flow in the Cruces River had been overestimated because Celco Arauco staffers used a questionable methodology. In addition, Zaror opined, the last water monitoring station in the river produced inaccurate estimates of water flow because it recorded ocean tides. Added together, these discrepancies consistently underestimated pollution and overestimated the river's capacity to dilute pollution.

This questioning of the data seemed to have no bounds. During the trial against Celco Arauco, scientists questioned how many swans had died, whether or not they simply migrated following seasonal patterns, and the value of ecosystem change. Some scientists described the swans as rapacious scavengers that had overforaged their way to starvation and tried to turn the disappearance of the luchecillo into a blessing:

> Luchecillo is an invasive species that needs to be eliminated. The black-neck[ed] swan is an opportunistic species that eats whatever it can find. (*CDE v. Arauco* 2013, 3503)

> Luchecillo is a submerged species, making it a highly aggressive weed that interrupts navigation. Its disappearance is equivalent to all the mice of a city disappearing. (Ibid., 3505)

Alternative hypotheses became increasingly complex and even philosophical. In court, a scientist from the University of Chile testified that environmental change was so pervasive that events at the Cruces River were trivial in comparison; climate change, natural disasters, drought, and ecological succession were the important things to study and understand.

This emphasis on the limits of scientific knowledge was picked up widely. For example, communications expert Eugenio Tironi (2011), whose clients include Celco Arauco, argued that the Cruces River crisis could not be explained through science because it is impossible to know what really happened there. And it ultimately undermined Austral scientists' confidence in their own knowledge of the wetland; in a book about these events written by some Austral scientists, they conclude, "Most likely, the limited knowledge we have of the wetland's [ecological] resiliency explains why it has been used and abused" (Escaida et al. 2014, 223).

One thing all scientists agreed on was praising peer review as the best method for validating scientific claims. The director of the Catholic University's ecologists, for instance, dismissed Austral's science because "when a hypothesis is only proposed in reports that are not circulated in the [peer-reviewed] literature, scientists cannot evaluate it" (*CDE v. Arauco* 2013, 3696–3697). Faced with such uncertainty around the credibility of monitoring data, the proliferation of alternative hypotheses, and efforts to redefine the value of the affected species, in interviews most scientists regarded peer review as the best guarantee of their independence and credibility.[15] Industry staff also valued peer review, and seemingly encouraged scientists to publish their Valdivia mill-related work in peer-reviewed journals to help legitimate it.[16] Some saw confirmation in the absence of peer-reviewed literature that Austral's science fulfilled a special interest for Conama.[17] Table 4.3 belies this, though: a count of peer-reviewed articles generated by this case shows that Austral scientists come out ahead of their rivals.

Table 4.3
Academic Publications Explaining the Ecological Crisis at the Carlos Anwandter Nature Sanctuary, Summary Data

Article	University first author	Peer review?	Reported funding sources	Causal element	Proof	Data sources	Area is described as
Pinochet et al. 2004	Austral	No	No info	Iron	Samples	Plant samples	Sanctuary
Ramirez et al. 2006	Austral	No	EULA	Ultraviolet radiation	Comparative case	Available data	*Bañados*
Mulsow and Grandjean 2006	Austral	No	Conicyt	Sulfur and acid	Lab experiment	New data	Sanctuary
Artacho et al. 2007a	Austral	Yes	Government	Iron	Samples	Swan samples	Sanctuary
Artacho et al. 2007b	Austral	Yes	Government	Iron	Samples	Swan samples	Sanctuary
Jaramillo et al. 2007	Austral	Yes	Government	Aluminum sulphate	Samples and comparative case	Available data	Sanctuary
Lovengreen et al. 2008	Austral	Yes	Austral	Reject ultraviolet radiation	Observation	New data	Sanctuary
Lagos et al. 2008	Austral	Yes	Government	Reject environmental causes	Samples and Landsat images	New data	Wetland
Nespolo et al. 2008	Austral	Yes	Government	Pollution	Samples	Swan samples	None
Palma et al. 2008	Catholic	Yes	Arauco	Reject pollution	Lab experiment	New data	River
Norambuena and Bozinovic 2009	Catholic	Yes	Arauco and Conicyt	Environmental causes	Samples	Swan samples	River
Marín et al. 2009	Chile	Yes	Government	Sediments and freezes	Observation and model	Available data	Wetland

A Bitter Rivalry Develops

The Catholic University's report raised doubts about the strength of the scientific evidence against Celco Arauco and also fractured the scientific community. Austral and Catholic University scientists became locked in a bitter rivalry for credibility. Their rivalry, and attempts to settle it through boundary work, illuminate debates about what counts as good and appropriate science in relationship to the state in light of the imagined relationship between the state and industry. Outside the Valdivia mill crisis, Austral and Catholic University scientists were close colleagues. The controversy, however, put into sharp relief the differences between them. While Austral scientists lived in the community and had intimate knowledge of the wetland, the Catholic University scientists lived in Santiago, 865 kilometers north of Valdivia. Instead of ready access to the wetland, they had easy access to Chile's political and business elite, many of whom had donated lavishly to the Catholic University, including its Center for Advanced Studies in Ecology and Biodiversity. The center's director, Fabián Jaksic, is a zoologist with a PhD from the University of California at Berkeley. Jaksic is close to the Christian Democratic Party and the like-minded think tank Centro de Estudios del Desarrollo. He has worked for academia and industry all over Chile. While Jaramillo dresses in woolly sweaters, Jaksic is more frequently seen in an elegant suit and tie. Jaksic is not shy about speaking in English, or touting his political and business connections.

Conama and Celco Arauco enrolled fundamentally different scientists to speak for them during the crisis, and years later, the state's public prosecutor and Celco Arauco called different expert witnesses to testify in the case for damages filed against the company (table 4.4). The first difference that stands out is financial: Celco Arauco called ninety-one experts, many of them from abroad, compared to the twenty-four that the state brought in. The witnesses differed also by place of work and residence as well as training and expertise. Austral scientists were mostly experts in flora, fauna, and water ecology, while the scientists that Celco Arauco called to the witness stand tended to be engineers and chemists. And they pursued divergent goals. Austral scientists were asked to find the causes of the swans' demise for Conama, while Catholic University scientists set out to raise doubts about this work.

Perhaps these differences reflected each funder's worldview or epistemic commitments. Alternatively, they may have resulted from a knowledge-based

Table 4.4
Witnesses Called to Declare in *CDE v. Arauco*

	Total	From Valdivia?	From abroad?	Areas of expertise	Organizations where they work
CDE	24	17	0	Mostly veterinarians and zoologists, and three chemists	Austral University and state agencies
Arauco	91	9	14	Mostly engineers and chemists along with a few economists and managers	Catholic University, Concepción University, EULA, consulting companies, and paper industry

strategy, pursued by the state or company to advance their legal and political goals (prosecution and defense, respectively). Or these divergent characteristics may have gained importance as the controversy unfolded, as scientists reacted to the bitter rivalry they found themselves in. In any case, these differences had consequences, not just for the making of local knowledge, as discussed earlier, but also for the values scientists brought to their work. For example, regarding scientific evidence, the scientists that testified for Celco Arauco had a relaxed attitude toward broken monitoring equipment and the discrepancies between the lab and monitoring reports handed to the authorities. Zaror called these "understandable mishaps" that take place when large industrial systems come online. By contrast, Austral scientists saw this as evidence of wrongdoing by Celco Arauco.

Scientists differed in what they called the area too, reflecting their relationships with the place. To locals, the area is above all the Carlos Anwandter Nature Sanctuary, a unique, protected space that defines the town of Valdivia. The sanctuary protects a wetland that was created in 1960 after nearly sixty square kilometers of land sank following a devastating, magnitude 9.6 earthquake—the strongest on record worldwide. Twenty years later, the wetland's unique geologic origin and biodiversity earned it a listing in the Ramsar Convention for Wetland Conservation. The sanctuary includes deeper rivers, like the Cruces, and shallow *bañados*, each with their own ecology, geology, and biology. The Cruces River flows through the wetland, past Valdivia, and out to the Pacific Ocean at Corral Bay, where Spanish colonial rulers constructed a large fort. The river first turns into a wetland and then the wetland turns into an estuary.

Table 4.5

Word Count of Common Names Used to Describe the Carlos Anwandter Nature Sanctuary in Scientific Reports (Word Count in Parentheses)

Austral	Ramsar	Zaror	Catholic University	CEA (1993)
Sanctuary (762)	Sanctuary (76)	River (64)	Wetland (54)	Sanctuary (51)
River (537)	River (65)	Sanctuary (23)	River (47)	River (15)
Estuary (109)	Wetland (13)	Wetland (2)	Sanctuary (39)	Wetland (2)
Wetland (63)	Estuary (6)		Estuary (6)	

Local scientists most frequently referred to this place as a sanctuary. This includes references in the Austral Report, an older report by a local NGO called CEA, and also the report by Ramsar scientists, sent to investigate because of the site's global significance (table 4.5; see also table 4.3). Sanctuary was also the most commonly used term by residents, over generic terms like rivers or nature (Sepúlveda 2016, 297–298; Delgado et al. 2009). By contrast, Catholic University scientists and others from outside Austral more often used terms like river or wetland. Scientists unfamiliar with the wetland, in other words, were less likely to acknowledge the social and cultural value the place had acquired for locals as a protected sanctuary.

This variety of naming practices magnified the scientific uncertainties about the area. In environmental regulation, "bridging objects" are particularly useful because they are statements of fact, concept, or procedure that reflect as well as sustain shared understandings among scientists and regulators of how a policy will impact nature. In Japan, for example, Kenneth Wilkening (2004) argues that the bridging concept "imported sulfate from China is acidifying Japan" foreclosed ongoing controversy over the exact causes of acid rain and facilitated international cooperation. In the Valdivia mill case, a suitable bridging object could have been "wetlands that are estuaries are especially vulnerable to industrial pollution."[18] For this kind of claim to become a stable bridging object, a scientific community with shared goals and understandings is necessary, as they articulate, repeat, and substantiate the claim until it wins over the alternatives. In the Valdivia mill case, however, scientists seemed bent on preventing this kind of stabilization. Some scientists may have been acting strategically to defend the company or their own reputation from accusations of doing "bad" science beholden to a patron's interests. Others may have been responding to an institutional environment characterized by contested ideals of Chilean science along with the absence of spaces for routine encounters between scientists and officials. Intentions aside, absent such spaces, scientists and officials could not negotiate together what science can and cannot credibly speak to. Building relationships of trust and mutual respect is difficult in these conditions.

About a year after the crisis began, the Ecology Society of Chile and Biology Society of Chile held a joint conference with a special panel, titled The Role of Ecologists in Environmental Management in Light of the Cruces River Case, to reconcile the Austral and Catholic University teams. A second goal was to discuss what role scientists should play in environmental crises such as the one occurring in the Cruces River. By most accounts, the societies' panel did not achieve closure among scientists but instead revealed disagreements over the role of corporate funding in science and risks to science from politics. The panel's keynote speakers were there to defend the Catholic University and clarify what they believed were the proper boundaries of science in a public controversy. Jaksic explained that ecologists "should not be judges" in EIAs, leaving the decision to the authorities. EIA director Raúl Arteaga worried that when ecologists "cross the line of what is natural to them," they try to impose their opinions as immutable truths, becoming judge and jury. Celco Arauco's new environmental manager called for more monitoring, more scientific research, and more exchange between scientists and industry (Informe sobre el simposio 2005, 3–4). The panelists defended an ideal of science as limited and independent of politics as well as its funders. They claimed that in fulfilling its contract with Conama, Austral had "overstepped its boundaries," transforming scientific opinion into political judgment.

Yet as most scientists recalled the event, the scientific community reprimanded the Catholic University for "acting like a think tank"—that is, reanalyzing existing data, without collecting or generating any new data, to fulfill a political party's or company's goals. Political parties in Chile typically have an associated think tank (Silva 2009). This criticism thus recast Jaksic's team not as scientists working at the nation's premier ecology department but rather as consultants whose work was unfit for the peer-reviewed literature.[19] Other statements reflected a conviction that the power of science in public affairs depends on clearly differentiating it from industry. Thus, some audience members vindicated what ecologists can contribute to collective decisions as ecologists, not as consultants, and several voiced a common complaint against Conama: the agency approved the mill for "political" reasons—that is, pressured by an executive government committed to exports and jobs—despite its better scientific judgment (Informe sobre el simposio 2005, 4–5). By contrast, Jaksic dismissed the idea that scientists cannot criticize their colleagues without new data as "absurd" and questioned the accusation that industry funding erodes scientific independence. Rather than isolate science, he argued, better to manage conflicts of interests with contracts and ethical codes of conduct to avoid any scientific "self-censorship." The special panel not only failed to reconcile the scientific community; it revealed in public a schism between those willing to blur the boundaries separating science from consulting, like Jaksic, and those anxious to demarcate them.

Jaramillo and Jaksic made their peace years later, in private, at another conference. Speaking alone as peers, they agreed they had been locked in an asymmetrical pursuit. The Austral University's team could not compete with the Catholic University's, given the latter's access to far greater funding. Austral scientists felt this was confirmed by the fact that thereafter, Chile's national science agency rejected their proposals on wetland or swan research because they "lacked expertise," posed questions that were "too specific," or "lacked historical data" to merit funding.[20] Public debates aside, in private scientists tended to agree that a money-inflected hierarchy exists. By contrast to Catholic University ecologists, whose center continued to have access to generous funding, Austral scientists felt financially punished for years afterward for the role they played in the controversy.

Jaramillo, meanwhile, began a personal project: each day as he drove to and from work, he counted the number of swans he saw from his car and kept a record of the wetland's gradual recovery. This was a zero-budget census of uncertain scientific value. It is a reminder that Jaramillo, like his Austral colleagues, continued to live where he worked and considered the place a sanctuary, not just a wetland.[21]

Science on Trial

In 2005, the state prosecutor sued Celco Arauco in the civil law courts of Valdivia for environmental damages. In 2008, the judge began to call witnesses to testify (table 4.4). Each witness had to undergo a vetting process to determine if they held a conflict of interest: was a scientist paid by Celco Arauco negatively predisposed to the state, and was a state-paid scientist biased against Celco Arauco? Courts of law have their own methods for determining the credibility of science, separate from those used by the public or officials (Jasanoff 1995). The trial provides another window into how Chilean scientists handled conflicts of interests and defined good science in a particularly understudied venue: courts of law. The transcripts from these vetting procedures and scientists' testimony show that scientists established their credibility by reifying science into the "scientific method" that produces probabilities, not truths, thus further undermining their abilities to speak assertively about nature. Instead of confident and informed, they appeared doubtful and defensive. Though the judge did not disqualify any scientist, neither did she rely much on their testimony. In 2011, she ordered seven new experts to review the evidence and submit individual assessments; six of the seven found Celco Arauco guilty, and in July 2013, the judge issued the corresponding verdict.

Celco Arauco's scientific witnesses answered questions about conflicts of interest in multiple ways, ranging from mundane to elaborate explanations. Pleading ignorance was probably the most common strategy: the subdirector of the Catholic University's ecology institute declared that he "supposes that there is some funding [from Celco Arauco] to the university but [he] does not know anything about it" (*CDE v. Arauco* 2013, 3971). At other times, however, scientists made convoluted arguments that involved affirming their independence by reaching results contrary to their funders' interests.[22] In this statement, a natural scientist from the University of Concepción who testified for Celco Arauco defended his credibility by saying,

> In many cases our results have been opposed to or contrary to the opinions of those who hire us. For example, one of the funders of our institutional program Water Quality Monitoring of the Bío-Bío River … is Endesa [an energy company]. I have conducted research published in international scientific journals like *Bio-diversity* and *Conservation* where I have publicly shown the negative impacts that hydroelectric generation produces in the Laja Lake. … To summarize, despite having received payment from [Celco Arauco], this does not imply a dependence for the results of the study. (Ibid., 1925)

Celco Arauco's lawyers similarly contended that "the independence of the [Catholic University's Ecology Institute] is not under discussion. … The researchers of the two most important universities of this country [the

Catholic University and University of Chile] have given great demonstrations of their independence with respect to public and private entities that contract research" (ibid., 3790). The lawyers then cited examples of research institutes that, in their view, work regularly for government and corporate entities yet can be presumed independent because they reach results contrary to their patron's interests. From the Catholic University, the examples were Dictuc, an applied consulting arm of the engineering department that works for the Ministry of Public Works, and the economics department, which in their view, issues recommendations contrary to the interests of its private funders. From the University of Chile, the lawyers said the "best example" is that of Victor Marín, an ecologist who was hired by Conaf, the national forestry agency, to study the Cruces River wetland. Marín's wetland study became controversial after Conaf refused to pay him, arguing that Marín's study was about the causes of the wetland's environmental collapse rather than measures to remediate it, which was what the agency had asked for. Nevertheless, Marín published this research in a peer-reviewed journal article (table 4.3). Celco Arauco's lawyers maintained that the fact that Marín's research had passed peer review was evidence that the state, not industry, manipulates scientific truths for its own ends. In other words, it is the state and publicly funded science that are suspect.

Jaramillo was subjected to an intense vetting process in which Celco Arauco's lawyers contended he was unreliable because he had a vested interest in proving the company's responsibility. They recast normal scientific activities, like presenting one's work to others, as a suspicious practice and questioned the reliability of science produced to pursue previously identified ends:

> [Jaramillo] is so invested in the "cause against [Arauco]"...that a verdict that does not recognize the role of [Arauco]...would cause the witness serious academic harm, hurting his capacity to earn his own living. Such a verdict would hurt his professional pride, as illustrated...when the witness said he would conduct research to prove wrong the conclusions reached by the College of Science of the University of Chile....This is itself a nonscientific attitude because the very essence of a scientific conclusion is to be merely provisional and probabilistic, so that scientists must maintain a prudential attitude and not commit spiritually or emotionally with hypothesis that must be open to discussion....These emotional factors have led him to commit to a crusade to transform his hypothesis into formal truth that by virtue of repetition he hopes to promote as true. (Ibid., 689)

This adversarial process strengthened a fundamental problem in Chilean science: the conflation of quality and credibility with money. Few scientists grasped the importance of money better than Jaksic, who then had more resources at his disposal than likely any other Chilean ecologist. In closing his testimony in favor of Celco Arauco, Jaksic said, "To be fair, funding for [research on chemical hypothesis] has been scarce. Nonetheless, I have to emphasize that long and expensive experiments like our mesocosm study [Palma et al. 2008] have rejected the chemical hypothesis at least as it pertains to the effluent as a whole. The quantity and quality of the experiments and analysis that promote natural hypothesis are higher" (*CDE v. Arauco* 2013, 3706). Jaksic thus expressed the ideal that good science is expensive science. In the cross-examination, though, he recognized that the mesocosm experiment had limited validity because it did not test the same effluent the mill produced when the disaster occurred. In the cross-examination, Jaksic explained, "[A test of original 2004 effluents] has not been asked for" (ibid., 3720).

The judge did not disqualify any scientist because, legally speaking, a conflict of interest requires direct pecuniary gain (ibid., verdict article 9). Yet for those suspicious of the private sector, signs of Celco Arauco's financial influence were everywhere. Celco Arauco's scientists presented evidence that was expensive to obtain or produce: satellite images, isotope tests, and sampling campaigns that government-paid scientists could never afford. Austral's original report, for instance, contained results from 5 sediment samples, while consultants working for the company tested 139 samples with much more sophisticated methods. On the witness stand, Austral's scientist in charge of sediments was blunt: if in 2005 they had had US$250,000 more, they could have analyzed sediment records to challenge Celco Arauco's claims.

The most common defense of science scientists used was to appeal to traditional methods, like peer review and narrow descriptions of the scientific method. An Austral scientist claimed that one positive result of this crisis was that his faith in the scientific method had been renewed: "I can now be confident that anywhere I go and anything I do I can apply this same scientific method to study any phenomenon."[23] Even economists, who use a different scientific method from ecologists, testified (on behalf of Celco Arauco) that "naturally [as an economist] I am not an expert on ecosystems, but whatever the discipline where the scientific method is applied, this conceptual framework [of hypothesis testing] must be present" (ibid., 994).

In court, Jaksic took this faith in the scientific method further to make claims about ecology as a discipline. He first dismissed Austral's work by setting up a hierarchy of hypotheses: "In environmental science, we consider hypotheses about natural change more parsimonious than one where change is caused by humans and their economic activities." He continued that the methods of ecology make it irrelevant to the study of pollution: "We only commit to study the biological, ecological, and biodiversity aspects for which we do not have to be informed about engineering issues like inputs or emissions that do not have to do with a biological evaluation of the state of the wetland and the changes that have occurred there." Moreover, he said, ecology and environmental science have nothing to say about harms resulting from pollution, which he defined solely in legal terms: "If the emission standard is met, it is not necessary to do any [scientific studies] of the environment because the regulator already decided under and above what thresholds we must worry about the environmental effects" (ibid., 3707–3708). With these statements, the director of Chile's most important ecology research institute argued that ecology and environmental science have no concept of anthropogenic environmental damage or toxicity.

These strategies ultimately erased scientific authority. Jaramillo, exhausted after an aggressive cross-examination, finally confessed, "I have previously sustained that science has no monopoly on the truth and therefore there are no absolute explanations." Another Austral scientist got locked into a tedious debate when Celco Arauco's lawyers asked him, "Do you share Edington's statement to the effect that scientific hypothesis should be signed with the phrase used by vendors, S.E.U.O, *save for error or omission*, because as is stated in Desiderio Papp's book, *Philosophy of Natural Science*, 'relativity' not only exists in physics but in all scientific conclusions?" The lawyers rambled on comparing trends versus snapshots, and reflecting on the inherent uncertainty of knowledge as illustrated by the revolutions caused by Albert Einstein and Stephen Hawking. The scientist replied that scientists must never cling to existing theories. Otherwise "as Dr. Maslow said, if you always use a hammer, all problems will begin to look like nails" (ibid., 2421). By all appearances, however, Chilean scientists transformed the scientific method into a hammer that they were reluctant to use on legal or socially important nails.

Attempting to demonstrate their credibility and independence, Chilean scientists put themselves in a catch-22, arguing that their credibility was evidenced by reaching results against their patron's interests, turning the scientific method into a blunt hammer-like tool, and marginalizing ecological expertise from the demonstration of environmental harms. These ideas were reaffirmed during interviews, with scientists repeating that science does not provide truths; at best it provides probabilities.[24] It is useful to briefly contrast Chilean scientists' experiences with those of their colleagues called by Celco Arauco to testify in Canadian, Swedish, Finish, and US courts. US judges asked things like, "What do you have to do to keep current that professional engineer's license?" "Are there any other specialties, awards, or publications you would like to describe?" and "How does your expertise relate to the opinions that you were asked to provide in this case?" These questions gave US witnesses the chance to explain why their expertise was relevant and what they contributed to the case. In contrast, Chilean experts spent much of their time disqualifying themselves, answering lawyers' questions with, "I am not a chemist," "I am not an agronomist," "I am not a forest engineer," "I am not a lawyer," and "I am not an enforcement agent."

Control Shifts from Conama to Celco Arauco

The first warning signs that something was wrong at the Valdivia mill appeared quickly; within two months of opening, the mill was fined and reprimanded for emitting foul-smelling sulfur air emissions. Within six months, scientists raised the alarm that the swan population was declining. Within seven months, a consulting agency hired by Conama reported that the Valdivia mill was built to a larger capacity than authorized (850,000 tons versus 550,000 tons per year), and would consume more water and emit more wastes than allowed by its EIA permit. The mill also had an unauthorized waste pipeline, had begun operating two weeks earlier than allowed, and was failing to meet several monitoring obligations. This report documented nineteen violations. Between 2004 and 2005, Celco Arauco was fined a total of almost US$300,000 for these and other violations.[25] In early 2005, Conama shut down the mill's operations, but soon allowed it to reopen—the first time in Chilean history that the state stopped industrial activity due to regulatory violations (Escaida et al. 2014; Sepúlveda 2016). In June 2005, Conama imposed tougher emission targets on the mill and required that the mill find another place to dump the waste effluent.[26]

Throughout 2004 and 2005, activists and legislators accused Conama of acting erratically. Some legislators said the agency had to be held responsible for approving the mill and then, belatedly, detecting irregularities.

Conama's director defended the agency, saying mistakes had been made given regulators' lack of experience, but they had done their best (Escaida et al. 2014, 133). Whatever the case, the Valdivia mill's permit illustrates that Conama did not have much of a plan in place to deal with emergencies. The permit read, "If unforeseen environmental impacts should arise, the company must inform … Conama …, and execute the necessary actions to mitigate, repair, or compensate these, as corresponds. The authorities must be informed immediately following the detection of environmental impacts" (Conama 1998, 71).

As discussed earlier, Conama's weaknesses stemmed from a lack of experience, legal limits to its authority, and the power of industry. At the time of the crisis, politicians and officials even questioned whether Conama had the authority to shut the mill for violations to its EIA permit (Sepúlveda 2016, 200).[27] Moreover, Conama's weak position was aggravated by a mutual lack of trust and respect between the agency and scientists. For instance, Conama requested three scientific reports, but did not have a plan for how to use this information or handle uncertain and countervailing conclusions. Overwhelmed with information, the agency lost confidence in the Austral report (the most comprehensive of the three it commissioned) and its own authority to act. Conama's loss of confidence was compounded by criticisms it deserved, such as how it had evaluated the mill's EIA studies or its flip-flopping on the monitoring scheme. These criticisms aimed to improve how Conama evaluated EIAs and enforced environmental regulations.

Occasionally, however, scientists launched criticisms so out of proportion with existing resources that they distracted from the direct causes of Conama's weaknesses, such as too few staff with too little in-house expertise and restricted rights to interfere in companies' operations. Zaror, for example, advocated for using DNA analysis to detect pollution—a proposal that sounds like science fiction given Conama's resources at the time. The agency could not enforce even the limited monitoring plan in place, having already caved to pressure from Celco Arauco to roll back the original monitoring scheme. It also had no scientific knowledge base on which to recommend sentinel species for the Cruces River, or access to a trusted network of scientists it could rely on for advice (significantly, Zaror does not specify how DNA testing would work in this case). Conama could not carry out surprise inspections of the Valdivia mill, and had to rely on information and effluent samples provided by the company. Pollution monitoring through

DNA samples was thus even less feasible than detecting toxicity through AOX, as Conama had proposed to do, which then could not be processed by labs in Chile. This kind of overblown criticism only deepened feelings of incompetence and uncertainty.

Yet another limit to Conama's authority came in the form of the legal definition of "pollution," codified into law in a way that gave lawmakers full control over future modifications of what counts as a pollutant or pollution. The 1994 national environmental framework law that created Conama and EIAs defines pollution as any substance that appears in concentrations *that exceed legal limits* (article 2, numeral c, law 19.300), as opposed to the more common legal standard that defines a pollutant as an unwanted or harmful substance.[28] This definition resulted from a legislative compromise needed to pass the environmental framework law through a hostile Senate. As one conservative senator explained, "[To define pollution as a harmful substance was] very perfect … from the scientific perspective of what a pollutant is. … [But] if we use this as the basis for legislation, we enter the world not of ecology but rather environmentalism" (LH 19.300, 451). Another senator argued,

> A polluting substance or element, in the literal sense, is everything that alters natural environmental conditions. For example, our breathing is, obviously, a form of pollution, and those who smoke double or triple it. But what does the Constitution guarantee? Not that we stop breathing or smoking, but the right to live in an environment free of pollution—that is, an environment that does not alter natural life conditions. What will signal those limits? The law. And when the limits are exceeded, we will be in the presence of pollution. (LH 19.300, 446)

In this context, the legal victory the state prosecutor won in July 2013 against Celco Arauco is important. In her ruling, the judge argued that although the Austral scientists could not rule out natural causes, they provided evidence that natural causes by themselves would not have produced the observed ecological collapse. The evidence pointed to the release of a high volume of wastes that more likely than not, produced a toxic reaction that would not have otherwise occurred. The state prosecutor built a successful case around the same science that Conama had, in April 2005, rejected as insufficiently convincing to close the mill.

After the worst of the crisis had passed, Conama got to work on revising emission and water quality standards for the Cruces River. Ten years later, in January 2015, the new standards regulating water quality in the Cruces

River were published. National standards regulating waste discharges into rivers or coastal waters have not been updated since 2001.[29] These delays are not unusual in Chilean environmental politics (OECD 2005). In the case of the Cruces River, the first proposal, made by Conama in 2006, was criticized by NGOs. The second proposal, based on studies by Austral scientists, stalled in a deliberative exercise.[30] Remediation of the wetland also stalled absent a guilty verdict.

Remediation finally began with the historic 2013 guilty verdict. The court required Celco Arauco to remediate the wetland in cooperation with the Social Scientific Council established for this purpose. This space, set up to be a "hybrid forum" (Callon, Lascoumes, and Barthe 2012) that provides diverse participants opportunities to discuss, vet, and negotiate scientific claims, has been configured as a representative—not an expert—body: the Environment Ministry has two representatives, the state prosecutor has three, other state agencies have seven in total, and NGOs have another three. The company itself has four representatives on the council—the largest minority block.

Celco Arauco's role on the Social Scientific Council is but one example of how it emerged a new company from the Cruces River disaster. The crisis prompted several executives to lose their jobs, and a new environment division was created. The director of this new environment division, Andrés Camaño, had experience managing conflicts in mining towns in Chile's north. He explained to me that in these mining towns, he learned that science can make any company's project "bulletproof" by producing what he called "research for corporate defense." In Camaño's model, each new paper and pulp mill is associated with a research consortium that transforms communities' concerns into a long-term research program. The consortium provides a structure for environmental monitoring and toxicity testing while avoiding the sources of distrust that, in Camaño's view, sustained the conflict at the Valdivia mill.[31] Unlike the Social Scientific Council, the research consortium is operated entirely by the company to forestall conflicts like that at Valdivia.

The consortium builds trust because it includes local universities—which are most trusted by communities—and foreign universities—which are most trusted by company executives. This structure allays public fears of conflicts of interest because, says Camaño, "it is unlikely that five or six universities would all be colluding with the company." It also allays fears among Celco Arauco executives that universities are opportunistic, agree-ing to do a project without the necessary capacity to execute it. This point underscores a common perception that Chilean universities are so desper-ate for funding, they will agree to any work. As Camaño said, monitoring is big business, and "if I offer you two or two and a half million [Chilean pesos], would you tell me you can't do it?"[32]

The first such consortium was created at the Nueva Aldea paper and pulp mill, built after the Cruces River disaster. Celco Arauco staff called this project the "anti-Mehuín" in reference to the fishing village that in the 1990s blocked Celco Arauco's access to their bay, sending the Valdivia mill's wastes to the sanctuary instead.[33] At Nueva Aldea, Celco Arauco employs what the company describes as "the best of the best" scientists and tech-nologies to produce valid scientific data and knowledge. The research ques-tions the consortium tackles came from public comments raised through the EIA's required public participation process. The resulting scientific data help to explain accidents and distribute responsibilities. The consortium also gives Celco Arauco executives various opportunities to interact with the community, such as by eating local fish and drinking local wine to dem-onstrate their faith in a safe environment. Since the Valdivia mill's crisis, and despite some episodes of environmental harm at several of its paper and pulp mills, Celco Arauco has avoided the kind of open conflict that occurred in Valdivia.[34]

The final thorn in Celco Arauco's side was its poor relationship with Aus-tral University. The conflict split the university, much as it did the scientific community at large, between those who believed that universities must cooperate with the private sector to have access to resources and those who took a principled stance against what they saw as corporate belligerence against scientific autonomy. Since 2005, Celco Arauco tried on multiple occasions to sign an agreement of cooperation with Austral University, only to face opposition from faculty, to the disappointment of those—including some scientists who were central in the conflict—who felt a rapproche-ment was necessary so scientists could gain access to industry resources and thus once again be relevant to society. The court's guilty verdict that mandated the creation of the Social Scientific Council broke this deadlock. To the company, embracing the verdict became an opportunity to heal the wounds with Austral University that had continued to fester almost a decade after the swans first fell out of the sky onto the streets of Valdivia.

Conclusion

The Valdivia mill conflict shows that when environmental crisis strikes, scientists are not generally considered authoritative or credible participants. Moreover, public engagement further erodes their credibility. The case illustrates a few trends that run counter to some of the findings and assumptions in the literature on special interest science. First, in Chile the suspicion that scientists hold conflicts of interest affected everyone: industry- and state-sponsored science both were suspect. Of course, not everyone embraced these ideas evenly or unambiguously; nevertheless, Celco Arauco's lawyers and scientists made these arguments in court, under oath, and their assertions were reproduced in the press and during interviews. Of particular importance, scientists themselves internalized the idea that public or private funding can shape the results of science. Second, in Chile the amount of funding drives legitimacy, so that good science is expensive science. In recognizing the significance of financial asymmetries to their work, Jaksic and Jaramillo confirmed at the microscale what many Chilean scientists hold to be true at the global scale: that far from universal, scientific practices are highly dependent on social and economic context. Relative poverty yields relative inferiority.

These beliefs shaped scientists' reactions to Austral's science about the ecological collapse. Although some scientists reprimanded the Catholic team for "acting like a think tank" and nearly all extolled peer review as central to the production of scientific proofs, to the end most non-Austral scientists believed in Celco Arauco's innocence, convinced that Austral had failed to produce sufficiently authoritative scientific proofs.[35] Skepticism of Austral was based on two main factors. First, many scientists believed that, as Celco Arauco's lawyers argued, Austral scientists were invested in the cause. This was painted as a professional investment—Jaramillo's reputation would suffer if Celco Arauco were found not guilty—as well as an emotional and political investment to side with a popular local cause. The suspicion therefore remained that Austral scientists were partial to their funder's (Conama's) interests. Worse, the case shows that Austral scientists did not have any ways to effectively perform their independence. The preferred strategy cited by their rivals—to "reach results contrary to their funders' interests"—only recognizes the power of funders to shape scientific results. In seeking to prove one's independence "despite the odds," this strategy leads to partial science all the same (Dahl 2008).

Second, scientists rejected Austral's standard of proof because it lacked a causal mechanism. It remained unclear exactly which substance, in what quantities, and under which climatic and ecological conditions had caused the purported toxic shock to the luchecillo. Absent this kind of detail, non-Austral scientists remained convinced that alternative explanations were plausible, whatever their shortcomings might be.

As a result of these suspicions and beliefs, publicly funded scientists were on the defensive as much or more than privately funded ones. Catholic University scientists effectively blurred the boundaries between scientists and consultants, while Austral scientists exemplified traditional struggles faced by Chilean scientists to prove their value to society. This all had negative consequences for the construction of local, place-based knowledge. After this enormous scientific effort put into studying the Valdivia wetland, the only consensus to emerge was around the unknowns.

Conama's inabilities and weaknesses, both in absolute terms and relative to Celco Arauco's capacities, provided fertile ground for this scientific rivalry. Conama acted erratically from the start; first rejecting the mill's EIA and then approving it, then twice closing the mill, but then rejecting Austral's accusations in favor of leaving the mill open for business. The law limited Conama's capacities: the agency could not suggest alternative locations or substantial modifications to the project; it could not require use of best-available technologies; it could not conduct unannounced inspections, or collect and analyze effluent samples; and it was unclear if it even had the legal authority to shut down the mill for environmental infractions. Hence the agency tried to construct its authority through the means available to it: by requesting studies from external experts. But without scientifically literate and experienced staff, Conama had a hard time managing multiple, competing scientific reports. Even before controversy erupted, when the agency drafted the mill's EIA permit, Conama's logic and technical foundations seemed confused and suspect.

By contrast, and in response to the Valdivia mill disaster, Celco Arauco developed a coherent strategy for making science credible. The company now sponsors "research consortia" at all its mills that combine the company's ideals of credible science with public demands for inclusivity and participation. And at the Valdivia mill, it participates in the court-mandated Social Scientific Council, also constructed as a deliberative space. In sharp contrast to Conama and other state agencies, Celco Arauco has a plan for managing the public credibility of science. This shift parallels that observed in the United States, where citizens living near refineries are taking their scientific claims about the environment not to state agencies but rather to industry, which is seen to have more resources and ability to respond to their demands (Ottinger 2013). In the Chilean or the North American case, however, will scientists have sufficient autonomy to raise an alarm or be critical of the company's actions, and if so, how will they signal their impartiality to society? What consequences will this have for pluralism in science, including that produced by citizens (ibid.)?

Before these new spaces existed, a common reaction in Chile among those concerned that Conama lacked the capacity to conduct fair and scientifically sound EIAs was to advocate for "more technical" EIAs. Many shared the idea that Conama had approved the Valdivia mill for "political" reasons—that is, despite having technical reasons to believe that the mill would produce terrible environmental damages, Conama was pressured into it by the executive government. For some, a more technical EIA meant one that relies on "better science," including more exhaustive and accurate baselines.[36] Yet others, like the residents of Mehuín, had a more radical response: skeptical that a Chilean state agency could ever produce a truly technical EIA, they blocked science, seeing it as a weak link in an otherwise-rigged evaluation process. Clearly, what might count as an "apolitical" EIA was (and remains) contested. The next chapter examines a conflict where activists deepened the weak link strategy, in that they attacked the accuracy and credibility of the EIA itself before the proposed mine was even under construction.

5   Conflict at the Pascua Lama Gold Mine

In 2001, huddled together in a room in Alto del Carmen, a village in the foothills of the Andes mountains, Sister Cristina and local councillor Luis Faura poured over the technical details of the Pascua Lama mine. The Pascua Lama project at the time proposed to remove 14.1 million ounces of gold, plus some silver and copper, from a site 5,000 meters above sea level on the border between Chile and Argentina (Barrick 2000, chapter 2). Sister Cristina and Faura were far below, at 674 meters, but they knew the mine would impact their lives through the rivers and roads that connected them to the mine, just 80 kilometers away (figure 5.1). Conama was responsible for evaluating environmental impacts based on scientific studies and public input collected as part of the project's EIA. During the required public participation process for Pascua Lama's first EIA, local residents had raised concerns about the mine's impacts on glaciers, among other things.

As Conama finished its evaluation of Pascua Lama's EIA, Sister Cristina and Faura inspected the EIA documents to assess whether Conama and the company, Barrick Gold, had addressed the community's concerns. Could Conama ensure that Pascua Lama would not acidify the water or cause the rivers to dry up? Could it guarantee that cyanide-laden trucks would not overturn as they sputtered up the steep country road, along the river and past the school? The EIA was supposed to provide evidence that these concerns had been studied, and that adequate measures had been introduced to either avoid harmful impacts or compensate the community for them. EIA documents in hand, Sister Cristina and Faura marked good answers to their concerns in green, representing hope; adequate answers in yellow; and inadequate answers in red, representing danger. After a day's work, red dominated their pages.

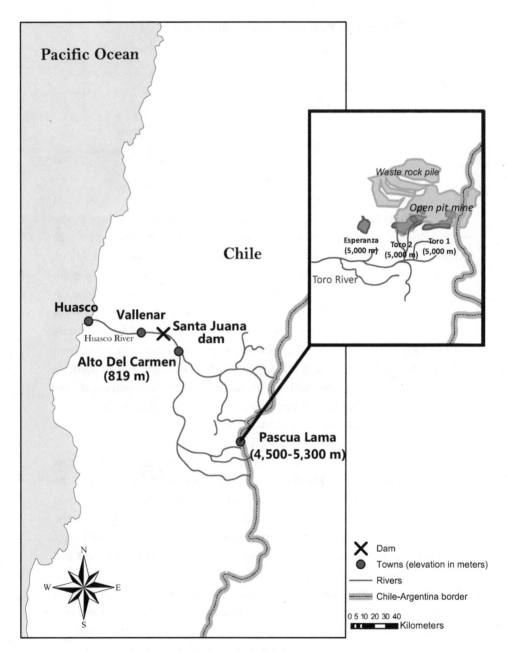

Figure 5.1
Map of the Huasco Valley and the Pascua Lama mine (insert).

Yet Conama (2001, 42–43) approved Pascua Lama's EIA. Most contro-versially, the agency approved Barrick's plan to "relocate" three small gla-ciers, "about ten hectares of ice," away from the mine's planned open pit to somewhere close by. Conama argued the ice could be preserved if it were moved somewhere morphologically similar to the glaciers' original loca-tion. If these measures failed to preserve the ice, then Barrick would have to maintain downstream river flow, although Conama left the details of how to achieve this vague. Despite having the permit it needed to build Pascua Lama, Barrick delayed the project as global gold prices fell. In 2004, after prices recovered and the company discovered more gold—three mil-lion ounces more—it submitted a second EIA to Conama to build an even larger mine at Pascua Lama, with greater environmental impacts. Regarding glaciers, this second EIA repeated the same information contained in the earlier project; Barrick estimated that roughly the same ten hectares needed to be removed. Barrick (2004) thus asked Conama to renew its approval of the same glacier relocation plan in this second EIA.

Unfortunately for Barrick, it had waited too long. Between Pascua Lama's first and second EIA, opposition to the mine had grown from a local concern to a national movement to protect glaciers from mining activities. Thanks to the pressure applied by activists in 2006, when Conama approved Pascua Lama's second EIA Barrick was required to redraw the pit's borders to leave the three small glaciers untouched, thereby foregoing access to about 7 percent of the gold deposits.

This chapter analyzes the Pascua Lama mine as a case of "slow violence," defined as "delayed destruction that is dispersed across time and space" (Nixon 2013, 2). Slow violence is relatively invisible, and poses represen-tational, narrative, and strategic challenges. As opposed to sudden, acute shocks—such as those examined in the previous two chapters—glaciers disappear slowly, gradually advancing the desertification of a valley. Gla-ciers have been traditionally managed as "water towers" that accumulate snow and ice during winter, and release freshwater during summer and periods of drought (Huggel et al. 2015). From the Andes to the Arctic, gla-ciers are increasingly vulnerable to climate change and mining. As glaciers recede and extraction technologies improve, mining companies are pursu-ing mineral deposits that were previously off-limits, thereby aggravating anthropogenic threats to glaciers (Bury 2015). Aware of these concerns, Pas-cua Lama's investors touted the mine as a model for a new generation of high-altitude extraction (Luna, Padilla, and Alcayaga 2004). To farmers in the Huasco Valley, however, Pascua Lama represented just the opposite: a direct threat to their agrarian way of life.

Narrating and representing episodes of slow violence is difficult, and glaciers—which are located far from most human's view, and to the untrained eye look like snow and appear static—pose particular challenges (Taillant 2015). The Pascua Lama project glaringly exposed these; many Chileans believed (including Conama officials involved in the second EIA) that Conama approved the first EIA without even knowing that the mine threatened glaciers. In other words, the state was blind to glaciers and their ecological importance. In a context where glaciers were politically invisible, activists turned to glaciologists for an independent scientific opinion that might legitimate their fears to a broader public that was also unaware of glaciers. Due to the activists' actions, at least seven glaciologists participated in the 2004–2006 EIA, compared to none in the 2000–2001 assessment. Trained in the study of ice, glaciologists have privileged methods for representing glaciers and, potentially, narrating the slow violence that thawing ice can produce.

This chapter traces how scientists participated in the Pascua Lama mine conflict and, in particular, when and how they were able to call into question claims made by other participants during the dispute. Glaciologists first helped to legitimate activists' claims, but then became increasingly marginal, even as the volume of scientific evidence grew. The first part of the chapter expands on the Pascua Lama project as a case of slow violence that exacerbates ecological vulnerabilities and leads to a resurgent "environmentalism of the poor" (Nixon 2013, 4; Martínez-Alier 2002). Huasco Valley farmers, poor and nonpoor alike, saw in Pascua Lama a threat not just to glaciers but also to their economic livelihoods. Their goal was to both vindicate the inherent value of water and protect their farms from an industrial threat. The case thus challenged Chilean elites' development rhetoric and a basic premise of EIAs: that the goal is to balance competing environmental and economic goals. Instead, every environment is important to someone's economic goals; thus, more accurately, EIAs balance different groups' competing economic goals. The second half of the chapter focuses on glaciologists and Pascua Lama's second EIA evaluation. Against expectations, Conama not only reversed course to protect the threatened glaciers but did so by introducing a potentially powerful concept: the precautionary principle.

Contradictions and logical jumps in Conama's rationale, however, rendered this decision less forceful than it might have been.

Glaciers and the Environmentalism of the Poor

The plans for Pascua Lama were exceptionally complex; once built, the open pit would be 2 kilometers wide and almost 1 kilometer deep, straddling the Chile-Argentina border, at between 3,800 and 5,400 meters above sea level.[1] At this altitude, oxygen is scarce, winds are ferocious, and temperatures are permanently below freezing. The working conditions are treacherous as well. These qualities exacerbate the environmental risks of any mine, but particularly those for one as large as Pascua Lama aspired to be. Of all the risks, those to water stood out because the Huasco Valley is both extremely arid and traditionally agricultural. The Huasco Valley marks an ecological boundary between the Atacama Desert and Chile's Mediterranean central valley. Families have for centuries practiced agriculture there, mostly for fruits and grapes. Some of these families belonged to Chile's traditional elite and own large farms that produce for export. But most are poor farmers that produce for local markets or personal consumption. The valley's geography

Table 5.1
Timeline of Events

1990s	Barrick acquires the site; mining explorations had been ongoing for some years and had likely already negatively impacted the glaciers in the area
2001	Pascua Lama's first EIA is approved by Conama, granting Barrick permission to move three glaciers—Toro 1, Toro 2, and Esperanza—to "preserve" them
2002	Opposition grows; activists involved include Sister Cristina, Luis Faura, Manuel Ossa, OLCA, and Diaguita indigenous communities; Ossa is invited to visit the Pascua Lama mine with two glaciologists (Gonzalo Barcaza and Pablo Wainstein)
2004	Barrick initiates studies for Pascua Lama's second EIA; Pedro Bazán at Conama becomes lead evaluator
2005	Opposition becomes national; the Huasco Valley's Irrigation Association joins the opposition; among other things, it hires EcoNorte to challenge the scientific and technical foundations of Barrick's EIA
	In June, however, Barrick and the Irrigation Association sign an agreement that allows Pascua Lama to move forward
2006	Conama approves Pascua Lama's second EIA, but denies Barrick permission to move the glaciers

reflects this social stratification: poorer farmers work upstream, closer to the proposed Pascua Lama site, while wealthier industrial farms are located downstream, where rivers carry more water (Campisi 2008).

Sister Cristina's congregation included the valley's poor. She initially welcomed Pascua Lama, believing Barrick's promises that Pascua Lama would bring jobs, improved irrigation canals, and safer roads. When Sister Cristina's church heard that Barrick said it would hire young men from the valley to work at the mine, the congregation sent the men to Santiago to receive training. But those jobs were never offered to the valley's youths, and the new infrastructure was never built. Sister Cristina and councillor Faura pointed to a string of broken promises. Well before they filled the EIA's pages with red ink, they had lost trust in Barrick and the Chilean state's willingness to serve the needs of the working poor. Moreover, the incipient activists were skeptical about Barrick's plans for the glaciers. They had seen the glaciers and knew them to be important. Through the 1990s, Sister Cristina's church offered mass every month to workers at Pascua Lama's base camp. The workers there showed the clerics satellite images of the glaciers. These images resonated with what the congregation had told her; the "perpetual banks" at the top of the Andes kept the Huasco River and its tributaries flowing, even through long droughts like that of the 1960s.[2] Thus, to Huasco residents and activists, Pascua Lama was always about economic well-being, first as a potential boon and later as likely a threat (Urkidi 2010).

The environmentalist alliance against Pascua Lama started at a spiritual retreat where Sister Cristina and councillor Faura met Manuel Ossa, a soft-spoken man from the Diego de Medellin Ecumenical Center, a religious organization that remains rooted in liberation theology and committed to social justice. Ossa then contacted the Latin American Observatory for Environmental Conflicts (OLCA), an NGO that supports communities concerned about their environment. Together, this alliance developed a message focused on the linkages between ecology and small-scale farming, in ways that invoke an environmental movement of the poor, where residents' "defense of Nature is based on a material interest in the environment as a source and a requirement for livelihood" (Martínez-Alier 2002, 11). In an open letter published in the Diego de Medellin Ecumenical Center's newsletter, Ossa (2001) denounced state authorities' willingness to approve projects for business reasons, and despite what the community or scientists might have to say, people in the [Huasco] Valley feel that Conama staffers who should be technical are political; and that in the area there are no experts in risk prevention or pollution with sufficient expertise for the Pascua Lama project. Indifference and hopelessness fill the valley, which is losing its youth. Many factors contribute to this hopelessness: a policy of "completed actions" practiced by Barrick; official submission to multinational companies; the suspicion that biased expert reports are well paid, and that political campaigns are financed by mining companies; [and] the feeling that the environment is subordinated to the interest of capital to maximize its profits, without the state applying any effective regulation. The lack of regulation is evident in official actions, some in the Huasco Valley, like that by former President Eduardo Frei, who said "national industry cannot be stopped" to avoid air pollution from petroleum coke burning in Huasco. This is the same spirit behind unprincipled and dirty policies in other parts of the country, like at Ralco [a dam project approved despite strong opposition]. All justified by one economic and ideological explanation, as repetitive as it is false, that "a country with US$5,000 annual per capita income cannot afford to steward the environment, like Switzerland and countries with US$20,000 income."

In this letter, Ossa places Pascua Lama in a larger national context characterized by procedural and distributive injustices committed by the state against communities. In Ossa's view, these injustices result from the great power of the industry relative to the weakness of the state. He presses for an environmentalism of the poor at the national scale—one that rejects the common claim that Chile as a country was too poor to protect its natural environment. In his view, Pascua Lama, like other industrial projects, was approved for political reasons over the better judgment of technically informed state officials or experts. Ossa uses the phrase "a policy of completed actions" to accuse Conama and government authorities of collecting scientific studies as well as holding public participation meetings about Pascua Lama under false pretenses. Far from ensuring a fair process that protects the collective interest, the EIA provides cover for a rigged decision that favors private interests.

In this context, activists zeroed in on glaciers for strategic reasons. First, glaciers inspired public anger against Conama. Conama gave Barrick permission to move glaciers to protect them—something no one believed was possible. Citizens in Huasco and across the country saw this as an insult to people's intelligence. Quite simply, "a five-year-old can understand you can't move glaciers," an activist told me.[3] To many, Barrick was getting away with an outlandish proposal because Conama appeared to think citizens were stupid enough to believe glaciers could be moved to be protected.

Barrick engineers described the plan to relocate glaciers in simple terms: it posed a logistical challenge because of the quantity of ice, but otherwise was just like other common practices used by people in northern climates. Speaking to a North American Congress on Latin America reporter, consultant Jeffrey Schmok explained, "They [people in northern climates] take ice, move it in shovels to pick up snow from the roads and they pick that up and move it...and that's essentially what's going to happen here" (Ross 2005, 17). Barrick's plan treated glaciers as if they were chunks of ice rather than ecosystems; the company proposed to take about ten hectares of ice and deposit it on top of a nearby glacier, called Guanaco, at which point the transferred ice would, in theory, adhere to the glacier (ibid.). Appalled, many Chileans could intuit that glaciers are ecosystems, not chunks of ice, that exist in a place for ecological reasons that escape human control.

Second, activists recognized glaciers as a "weak link" in Pascua Lama's EIA. Early in their organizing against Pascua Lama, a leading activist realized that "a few [expensive and scientific] reports would not lead [Huasco residents] to abandon what they knew to be true from experience."[4] In other words, Barrick's efforts to answer questions about glaciers in the EIA would be expensive but unpersuasive: no amount of science would dissuade residents from their experiential knowledge of these perpetual banks. To zero in on glaciers, then, provided an ideal way for less powerful citizens to stall the regulatory process and raise Barrick's business costs. Like the activists who mobilized against Celco Arauco's waste pipeline in Mehuín, discussed in chapter 4, anti–Pascua Lama activists identified the production of scientific knowledge about a particular place as the best moment to disrupt the EIA.

These concerns with water, poverty, and procedural fairness came together one day in fall 2004 in Alto del Carmen's main square. That day, national dignitaries came to the village to make a historic announcement: the state would transfer seventy-four irrigation dams around the country to local control, starting with the Huasco Valley's Santa Juana dam. The Huasco Valley's Irrigation Association (Junta de Vigilancia de Regantes del Rio Huasco) would now buy the dam back from the state. Under the provisions of Chile's 1980 constitution and 1981 Water Code, rural communities are responsible for building and maintaining local irrigation infrastructure. In this model, the state builds the infrastructure, and then agricultural producers, organized in an irrigators' association, buy the infrastructure back from the state. That day in Alto del Carmen, the Huasco Valley's Irrigation

Association shared the stage with President Lagos, who was going to announce the sale of Santa Juana to the association for almost US$40 million.[5] But the announcement was quickly canceled when several activists indicated their intention to interrupt the event.

In Alto del Carmen's main square that day, several local youths accused Barrick, its accomplices in the government, and the Irrigation Association of "killing our valley." They were "assassins of glaciers" in a "valley [that] is dying."[6] The association president's own son implored his father to act: "What will you do to save us?" Opposition to Pascua Lama was coming from the local elite's own sons, not only fringe environmentalists and poor peasant farmers. Moreover, if Pascua Lama acidified or depleted the valley's waters, Santa Juana would become a financial burden. The Irrigation Association now had an economic reason as well as a moral obligation to scrutinize Barrick's proposed mine.

Irrigators' associations are powerful organizations in Chile's dry north; there, "you are no one outside the association," an aide explained to me.[7] But power within associations is unevenly distributed. Votes are allocated according to water rights, so large water users like industrial farmers and mining companies like Barrick hold the most votes. Once the Huasco Valley's association got involved, the fight for and against Pascua Lama became "a fight between Goliaths" (*una pelea entre grandes*). Huasco's concerned residents and young activists were elated to receive the irrigators' support; their different worldviews would surface only later.

Reactions to the Poor's Environmentalism

Staffers working for Barrick and Conama rejected activists' concerns in ways that point to ethical gulfs between these groups. Initially, Conama and Barrick defended Pascua Lama using the same terms activists had used to criticize it. In response to Ossa's letter, a Conama official sent a letter to the same magazine describing what he saw as a "hopeful future" for the Huasco Valley, thanks to investments in roads, energy, telephone lines, and drinking water that would follow from the construction of Pascua Lama (Alvarez 2001). In the same issue, a Barrick employee concurred:

> As a mining lawyer I know mining poses risks for the surrounding area and communities. But this can be mitigated or at least compensated. What cannot be mitigated without economic investments are the lack of education, of health

services, and employment opportunities that today hurt the [Huasco] Valley's residents. We are convinced that with our project, and with the necessary safeguards, the [Huasco] Valley will not "die" as you write, but will "resuscitate." (Deisley 2001)

Both Conama's and Barrick's letter writers defended the procedure used to approve Pascua Lama's first EIA. The Conama official explained that "professionals from the different government agencies with competence on these issues" had visited Pascua Lama and could attest the glaciers had not been damaged (Alvarez 2001). Barrick's lawyer touted the company's international reputation and compliance with local regulations as evidence of its competence. State officials and industry elites thus defended Pascua Lama and its EIA: the process was legitimate, and the mine would bring local communities economic benefits. In their view, there were no procedural or distributive injustices.

This confident narrative changed with Pascua Lama's second EIA and the rise of glaciers to national prominence. Industry, government, and scientific elites were taken aback by the outcry over glaciers. An environmental consultant who had worked on Pascua Lama since 2000 recalled that family and friends were suddenly asking "if this [glacier movement plan] wasn't going too far," and she did not know what to say.[8] In the 2010 documentary *El Tesoro de América*, Barrick's founder and president Peter Munk said he assumed people were upset because glaciers evoke the "pristine" and "pure." Another Barrick lawyer believed that "when people hear 'glacier,' they think of large majestic glaciers, like the Laguna San Rafael," an ice mass of more than sixty-seven hundred square miles in Southern Chile in a national park of the same name.[9] In contrast, he implied, the glaciers at Pascua Lama were tiny and unimportant. If people could see them, they would stop worrying about their possible destruction. Scientists, too, were disinclined to champion activists' arguments. One scientist complained that what activists saw as "pristine, pure, untouched waters" were really "a toxic stew of acidity and metals."[10] In short, as the conflict became national, elites began dismissing activists' concerns as if they were advocating for romantic environmentalist ideals—what Joan Martínez-Alier (2002, 1) has called "cult of wilderness" environmentalism.

As the fight for and against Pascua Lama intensified, consultants, scientists, and staff members at Barrick misconstrued activists' concerns with glaciers as a cult of wilderness attitude. For example, in assuming that activists cared about glaciers because they are pristine, Barrick's Munk refused to recognize communities' economic concerns, grounded in their fear that his mining project would render their farms economically and ecologically unsustainable. In another instance, if Huasco farmers listened to scientists who told them the waters in Huasco's rivers were a toxic stew, then they would be forced to abandon any hope that the state could step in to supply them with clean water for their crops or homes. In Martínez-Alier's (ibid., 11) typology, each form of environmental concern reflects different values; while the cult of wilderness is ethically grounded in the protection of landscapes, the environmentalism of the poor speaks to "a demand for contemporary social justice among humans." Although Pascua Lama brought national attention to environmentalism as an ethic of social justice, these varied reactions to activists' concerns with glaciers demonstrate that elites in government, science, and industry did not share the same ethic as the activists.

Throughout both EIAs, Conama staffers claimed to be ignorant or were seen as such by everyone else. Pascua Lama's first EIA permit justified relocating the glaciers because they represented just 3 percent of all glaciers in the area (Kronenberg 2013). A Conama evaluator involved in this decision confirmed that this was their rationale; he recalled that staff at the agency considered the glaciers at Pascua Lama insignificant.[11] But nearly everyone took exception to this explanation, and instead understood Conama's decision to be the result of incompetence or indifference. A scientist reported a conversation in which someone said, "Barrick kicked the ball, and if the state didn't catch it, it's a goal."[12] Another glaciologist felt Conama's decision reflected a national cultural deficit: "Chileans do not have a mountain culture."[13] Others pointed to a gap in state capacity: "The state cannot protect anything if it does not know what is there."[14] Some Conama employees recognized the limits of their knowledge. In the first EIA, they were just "learning" how to evaluate such things; during the second, they still "had no knowledge of what glaciers meant" (Bórquez 2007, 69, 72; Taillant 2015).

This idea that Conama was just starting to build the capacity to know about glaciers was emphasized by Pedro Bazán, one of the agency's lead evaluators on Pascua Lama's second EIA.[15] Young and energetic, Bazán worked on Pascua Lama with enthusiasm; this was his first big project, and hence an opportunity to prove himself and the legitimacy of Conama's procedures. In his recollection, the second evaluation began as if from a clean slate: "We really knew nothing about glaciers. It was a completely new topic for us." He believed no one had raised concerns with glaciers during Pascua Lama's first EIA (which is incorrect as reflected in public EIA documents), and explained that the second time around, the evaluators became suspicious because the mine had expanded, but the anticipated impacts on glaciers had not. This prompted Conama, "in all innocence," to demand more information from Barrick. The public outcry took them by surprise.[16]

Together these responses deepened citizens' distrust of the EIA, Conama, Barrick, and a democratic regime committed to the needs of industry over those of rural communities. The real or perceived demonstrations of Conama's incompetence speak directly to concerns that procedural injustices are endemic to how Conama evaluates EIAs (Bórquez 2007, 71). Conama's incompetence only fueled the public's suspicions that EIAs are based on "political" rather than "technical" justifications, reflecting a policy of "completed actions" where political and business interests go unchecked. Additionally, in misrepresenting activists' concerns as if they were based on romantic idealizations of nature, elites deepened activists' suspicion that another act of distributive injustice was about to take place. Not only were citizens' interests procedurally marginalized; their voices also were misrepresented in ways that obscured their economic concerns. In this case, procedural and distributive justice concerns negatively reinforced each other, as distrust of the process fueled skepticism of its benefits (Urkidi and Walter 2011). In this context, activists turned to glaciologists for help, putting scientists at the center of a dispute that confronted competing economic, environmental, and ethical interests.

Selecting Glaciologists to Work With

Because sometimes "seeing is believing," Barrick invited Ossa to visit Pascua Lama. If people could see the glaciers, Barrick staff thought, their fears would be put to rest. Ossa and his environmentalist allies searched for glaciologists who might accompany him. The activists believed that glaciologists could provide an impartial validation of their claim that the proposal to move the glaciers would only destroy them. While activists turned to science to help legitimate their claims against the state and expose the state's lack of capacity, Barrick instead used the EIA to introduce concessions that might earn its project more support from evaluators and the general public. This was a lesson Barrick learned from the conflict at the Valdivia mill (chapter 4); the company might gain public trust if it could demonstrate its willingness to improve the project's technical features through the EIA (Jarur and Maldonado 2005). These different strategies reflect a trend in Chilean environmental politics to use the EIA as a site of both protest and negotiation.

Glaciologists got involved in Pascua Lama's second EIA evaluation at the invitation of each interested party—activists, Conama, Barrick, and the Irrigation Association (of which Barrick was a member). Reflecting their divergent environmental ethics and socioeconomic circumstances, each party used disparate criteria to select glaciologists to work with. Thus, the signals of scientific credibility were far from universal across different sectors of Chilean society. Activists like Ossa valued scientific impartiality, demonstrated by independence from the mining industry, as paramount. Finding such independent glaciologists proved incredibly difficult: Ossa asked friends at universities and consulting companies, but soon learned that most Chilean glaciologists work for mining companies. To the activists, any past experience with a corporation tainted scientists' credibility. Becoming increasingly desperate as he ran through his contacts, Ossa asked an NGO, Mining-Watch Canada, to send a glaciologist, but the organization lacked funding to fly a Canadian to Chile. A friend then put Ossa in touch with two young glaciologists—Gonzalo Barcaza and Pablo Wainstein. Ossa's friend had met the glaciologists by chance, after one of them rescued his child's stray ball in their apartment building. Barcaza and Wainstein were just finishing their university studies in glaciology; they were too young to have worked for anyone in industry and therefore met the activists' criteria for impartiality.

Barcaza and Wainstein accompanied Ossa on his visit to the mine site. Afterward, they published a short report enumerating the inaccuracies, gaps, and problems with Pascua Lama's first EIA. Given what they saw on the ground, they argued that moving the glaciers would destroy them and recommended compensating valley communities for their loss instead; this was the honest thing to do, they wrote (Barcaza and Wainstein 2002). From this point onward, more and more glaciologists got involved in Pascua Lama's EIA.

Barrick first hired glaciologists from a global consulting company called Golder, but the Irrigation Association then forced it to pay for additional glaciologists. Chilean glaciologists dismissed the scientist who Golder sent as an expert on ski slopes, not glaciers. As part of its negotiated agreement with Barrick (discussed in more detail below), the association required Barrick to pay for three additional glaciologists—two French and one Argentinean—selected by EcoNorte. EcoNorte was a small consulting outfit at the service of the Huasco Valley's Irrigation Association brought in to challenge Barrick's EIA in scientific terms. To compensate for its small size and ensure that its consultants could "speak as peers to Barrick's consultants," EcoNorte selected foreign glaciologists with stellar academic credentials.[17] The Frenchman Michel Vallon pioneered Andean glaciology in the 1970s, while the Argentinean Juan Pablo Milana was an authority on Chile's and Argentina's arid glaciers. The third glaciologist was Vallon's student, Christian Vincent.

State agencies like Conama and the Water Agency also mobilized glaciologists the second time around. Drawing a sharp distinction with the first EIA, Conama's Bazán felt it was important "not to go down in history as the only agency to approve a project about glaciers without the opinion of even one glaciologist."[18] Jumping into action, he obtained funding from Conama's central office in Santiago for two scientific studies: an improved baseline study and an analysis of the impacts of dust on the glaciers. He then proceeded to look for "any glaciologist" who could execute these. As far as he or anyone in Conama knew, there were no ice experts in Chile. When by chance he and his team found a university professor turned consultant with extensive experience in the mining industry, Bazán felt lucky. This combination of industrial and academic experience was "perfect" because it would appeal to legislators from all political parties.

In parallel, the Water Agency sent a team to Pascua Lama led by Fernando Escobar, an agency official who monitors glaciers around Santiago. In 2005, Escobar was just returning to glacier work after a hiatus he took in protest of the state's disregard for glaciers: the best demonstration of the state's ignorance, according to Escobar, was that despite his position as the state's only in-house expert on glaciers, he never heard of the Pascua Lama mine during the 2000–2001 period. Although Escobar has extensive field experience, his expertise is often overlooked because he does not have a PhD (Bórquez 2007, 72). Finally, in an unusual move, the regional Atacama government hired its own expert, Andrés Rivera of the Center for Scientific Research, an institute with a large glaciology group based in Valdivia, Chile.

The criteria that activists, industry, and state officials used to select glaciologists could hardly be more varied. While activists were distrustful of industry experience, state officials and legislators sought it out. Academic credentials—frequently assumed to be the most common indicator of credibility—only mattered to the business sector. Yet everyone overlooked the glaciology expertise of Chile's only experienced agency official because he lacked formal credentials. None of these patrons seemed particularly concerned about considerations related to glaciology as a science or practice. As a science, glaciology sits between disciplines, and is taught at schools of geography, geology, and engineering, where glaciologists receive different amounts of training in chemistry, biology, and ecology. Chilean glaciology began in the 1950s, when French and Japanese glaciologists arrived to study two fairly unique types of glaciers common in Chile: low-altitude glaciers, like those in Patagonia, and arid glaciers, like those in Huasco. Relative to other sciences, glaciology maintained itself through the 1970s and 1980s because the military had an interest in Antarctica, where Chile has land claims. Glaciology fieldwork is arduous, expensive, and done in teams. When choosing glaciologists, the interest groups involved in Pascua Lama never asked about field experience, or whether a candidate had a background in hydrology, chemistry, or ecology. Did they specialize in arid or wet glaciers, or did they have experience with both? With the exception of EcoNorte, none of the contracting parties gained a working knowledge of glaciology as a discipline and practice.

In contrast to their patrons, glaciologists had far more flexible identities, with their own markers of credibility. In private, glaciologists I interviewed were quick to dismiss each other as "opportunistic" or "uncooperative."[19] Opportunistic glaciologists changed their opinions too easily and appeared to be "too political," while uncooperative glaciologists had either "sold out to industry" or appeared fearful of getting embroiled in politics. All glaciologists ran the risk, according to their peers, of giving "impractical advice" such as restricting mining activity. Glaciologists were also flexible in who they worked for. All the glaciologists who worked on Pascua Lama at one point changed "sides": those who worked for Barrick later worked for the state, and vice versa. One result of this is that few glaciologists would meet activists' standards of credibility; if Ossa were to look for impartial glaciologists, he would again do best to look abroad or for students. Moreover, glaciologists did not share activists' social justice concerns, as evidenced by their quick dismissal of colleagues who appeared too political or gave impractical advice. The net result of these factors is that as scientists got involved, activists' concerns for procedural and distributive justice continued to be marginalized.

Faltering toward a More Just EIA?

Soon after Barrick submitted Pascua Lama's second EIA, the Huasco Valley's Irrigation Association mobilized against the project. Like the activists had done before it, the association's first step was to hire its own "independent, scientific-technical" assessment of Pascua Lama's EIA, through EcoNorte.[20] In February 2005, EcoNorte presented its appraisal of Pascua Lama to a large public meeting held in Alto del Carmen's main plaza. EcoNorte raised several issues of concern: Barrick's proposed methods for managing waste, avoiding acid rock drainage, and minimizing impacts on water could all be improved. Armed with this knowledge, for five months the Irrigation Association supported anti–Pascua Lama efforts. At that point, however, it joined Barrick at the negotiating table, only to reach an agreement that devastated activists (Junta 2005).

The Irrigation Association sat down to negotiate because its leadership felt it was in a bind. Not only was Barrick its business partner (Barrick was a member of the association and held the most water rights in the valley), but Pascua Lama also had an EIA permit from 2001 and could therefore legally operate. Association executives believed no company would walk away from so much gold; the project was, in their view, inevitable. Furthermore, they were under pressure from the executive government; an aide claimed that the association was told to "get the best deal it could get [from Barrick]" because the government's support for its position was limited.[21] In short, to the industrial farmers who then controlled the association, negotiating a solution with Barrick was the businesslike thing to do. This decision confirms and reflects the deep socioeconomic divides within the association; although many members are poor peasant farmers, they have been excluded from leadership positions within the association.[22]

Under the agreement, Barrick agreed to pay the Huasco Valley's Irrigation Association US$60 million over ten years for irrigation infrastructure and environmental improvements. These improvements included things like installing thirty online water-monitoring stations along the Huasco River and its tributaries. But the agreement left intact Barrick's original plan to move the glaciers. The association agreed to replace the water supplied by the glaciers with a new dam that could accumulate water from snowfall and release it slowly during summer months.

Barrick also conceded the rest of the EIA evaluation. From that point on, the association's consultants at EcoNorte would answer Conama's EIA-related questions on Barrick's behalf. With this switch, scientists who saw themselves as working for the irrigators—not Barrick—could transform the association's demands into legal requirements by introducing them into EIA documents. For example, there was no technical reason to require thirty water-monitoring stations; a lawyer proposed the number during the negotiation. With EcoNorte in charge, association scientists could craft a technical justification for a political agreement, making it all official through the EIA process. The switch in consultants thus reflected the business partners' efforts to legitimize their otherwise-extrajudicial agreement through science.

If the objective was to legitimate a political agreement through science, EcoNorte was an admittedly unusual choice. In contrast to multinational consulting groups like Golder, EcoNorte is a small firm based in Ovalle, a rural town just south of the Huasco Valley, and does a lot of work for irrigators' associations across northern Chile. Instead of tech-savvy engineers, EcoNorte's consultants were PhD scientists who had trained in universities in Germany, Cuba, and the United States, and settled in rural Chile for personal reasons. While association executives saw themselves as Barrick's peers, EcoNorte's staff members saw themselves as scientists more than consultants, as the "little guys" called in to solve the problems created by the "big guys." They had to fight to gain Barrick's respect: for months while they worked on Pascua Lama's second EIA, Barrick staffers tried to undermine them. Staff members refused to grant EcoNorte scientists access to data, saying, "You won't even have time to look at it." A Barrick executive wrote out a speech for EcoNorte's director in an attempt to undermine her authority. Another tactic involved questioning the need for those thirty monitoring stations, to which EcoNorte's scientist replied they could not touch what had been negotiated.[23]

EcoNorte faced yet another challenge to its authority: the poor quality of Pascua Lama's EIA materials. The first EIA had a bibliography that contained in total just forty-four references, most of them about flora and fauna, and none about glaciers (Barrick 2000, annex G, bibliography). The second EIA was slightly better. It included three aerial photographs of glaciers from different years (1986, 2001, and 2004, found in Barrick 2004, annex D), from which the consultants concluded that "the general size of the glaciers [did] not vary much" and that the glaciers were "compact ice … impossible to distinguish from snow" (ibid., chapter 5, 23–24).[24] Before EcoNorte took over the second EIA, Barrick's global consultants from Golder had responded to an initial round of questions from Conama by noting three "successful" examples of moving glaciers: a glacier conditioned into a ski slope in Canada, glaciers "moved" from around the Kumtor mine in Kyrgyzstan, and another plan that was abandoned as impractical.[25] These examples, all failures in different ways, only increased citizens' and scientists' distrust of Barrick's competence and willingness to manage Pascua Lama's impacts.[26]

In this context, EcoNorte's team turned to science to legitimate Barrick's plan, agreed to by the Irrigation Association, to remove and compensate for the three glaciers. The glaciers' relocation remained unpopular and needed to be publicly justified. EcoNorte's scientists proposed relocating the three affected glaciers, Toro 1, Toro 2, and Esperanza, within their watershed (figure 5.2). They justified this by recasting Pascua Lama as both unique and universal, and redefining the three glaciers as ice patches rather than traditional glaciers:

> The new Glacier Relocation Plan will not affect typical or traditional glaciers because the ice patches at issue are "reserve glaciers." They do not have any lateral movement except a likely undetectable reptation.…[27]
>
> Many international experiences with moving glaciers exist, but few if any involve ice patches with a similar metabolism, at similar altitude or conditions as these, as no places comparable to the dry Andes exist. The Relocation Plan faces no technical difficulties, except the possibility that the ice will be lost. Most likely, this ice will melt from natural causes following observed trends.…
>
> In sum, no project like Pascua Lama exists. In all likelihood some have existed, but never before was a detailed plan like this drawn up. Rather the ice was simply treated as waste rock. The lack of comparable experiences does not mean that useful lessons about glacier management cannot be extracted from other projects, because glacier management is a universal practice. (Barrick 2005b, chapter 3, 66–67)

This passage illustrates how the political negotiation between Barrick and the Irrigations Association required an intellectual negotiation, achieved by reclassifying the three threatened glaciers as reserve glaciers in which the top and bottom sections are homogeneous ice as opposed to differentiated zones of accumulation and loss. Louis Lliboutry, a French glaciologist who pioneered glaciology science in Chile with his 1956 treatise *Snow and Glaciers in Chile*, was the first to identify reserve glaciers. Whether on purpose

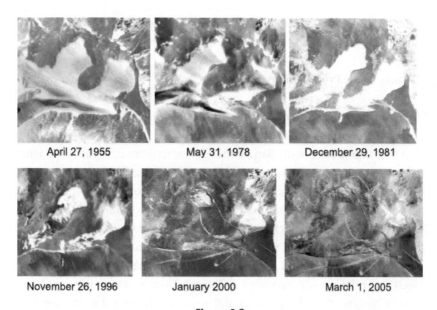

April 27, 1955   May 31, 1978   December 29, 1981

November 26, 1996   January 2000   March 1, 2005

Figure 5.2
Aerial images of Toro 1 and Toro 2 glaciers over time. Prepared by Juan Pablo
Milana for the report *Línea base de la criósfera, proyecto Pascua Lama* (2005).
Reproduced with permission from Milana.

or not, the Chilean roots of the term reinforced EcoNorte's assertion that
Pascua Lama was both unique and universal.

EcoNorte's argument for reserve glaciers convinced neither the public
nor Conama. As discussed in the next section, Conama approved Pascua
Lama on terms that violated the Barrick-irrigators agreement. Meanwhile,
activists were outraged by what they perceived as a downgrading of Toro 1,
Toro 2, and Esperanza from glaciers to reserve glaciers, ice patches, and—
the term most used in Spanish—*glaciaretes*, which sounds like a diminutive
of the term *glaciares*. To activists, this was a linguistic sleight of hand introduced by the company to absolve the government and industry of responsibility for the glaciers' health. This compounded the activists' pessimism
regarding the Barrick-irrigators agreement, which they saw as confirmation
of the policy of completed actions that Ossa had denounced. The agreement
demonstrated the divisive power of money. Barrick was not only buying
valley residents' support through social programs but also giving millions
of dollars directly to the valley's most important civic organization via the

agreement. As one activist told me, "Ours is a broken community," divided between those who refused and those who accepted Barrick's money.[28]

Getting to the Precautionary Principle

Back at Conama, state officials had to decide whether to accept or reject EcoNorte's arguments that these were glaciaretes whose loss could be made good through a dam, as the Barrick-irrigators agreement called for. If Conama rejected these contentions, agency officials could decide to protect the glaciers or require some other scheme of their choice to address the situation. Conama official and Pascua Lama evaluator Bazán explained that this time, "thankfully [Conama officials] had been armored from politics."[29] Instead, his team at Conama had made a supposedly technical decision to reject EcoNorte's arguments and protect the glaciers. His narrative and Conama's decision, however—as recorded in Pascua Lama's EIA permit, analyzed below—suggest that the political and technical are not shared categories across different sectors of Chilean society. More concretely, in this case state officials adopted surprisingly incoherent technical criteria, with consequences for how accountability would be practiced.

Bazán explained that throughout Pascua Lama's EIA evaluation, Conama officials like himself were at pains to distance themselves from the Barrick-irrigators agreement and also under pressure to protect the three glaciers. To demonstrate his distance from the politics of the case, Bazán claimed to have never even heard of the agreement—although this seems impossible given his position and the news coverage the agreement received. Nevertheless, as he recalled the events that led up to Conama's decision regarding Pascua Lama's second EIA, he was forced to recognize that politics did play a role; he understood that Conama had to approve Pascua Lama because "no one [in state agencies] would go against what the president of the republic says." "Luckily, this time technical and political criteria coincided," because the president would not accept the political costs of destroying the glaciers.[30] According to Bazán, a political and technical consensus to protect the glaciers emerged during a special meeting between Conama, the Water Agency, the Atacama regional government, and a senator for the Atacama region, with each government and state official accompanied by their own scientific adviser.

Bazán's description of the meeting suggests three estates were present that day: elected officials, representing political criteria; state officials working at state agencies, representing technical criteria; and scientists, who did not play a prominent role in Bazán's recollections. As reflected both in the studies they submitted at the time and in interviews, the glaciologists who advised all these parties during Pascua Lama's second EIA confidently considered Toro 1, Toro 2, and Esperanza to be reserve glaciers, glaciaretes, "glorified ice patches," or "relict ice." Whoever they worked for, all the hired scientists agreed that the glaciers were insignificant sources of water that could not be protected. Yet this scientific assessment was shunned by both Bazán as he recalled these events and Conama, as reflected in the EIA permit it issued to Barrick.

Rather than listen to the scientists, Conama adopted the precautionary principle to protect the Toro 1, Toro 2, and Esperanza glaciers. Conama made its case in two parts. First, the state now recognized glaciers as part of the community's "natural patrimony," thereby deserving of special protection. Second, Conama considered the projected water losses resulting from glacier removal to be uncertain. Regulatory agencies apply the precautionary principle precisely when scientific uncertainties surround potential harm; in these cases, agencies prefer to prevent possible harm while scientific studies continue to search for proof of harm. But Conama's otherwise-praiseworthy rationale soon encountered crippling contradictions that instead of protecting Chilean glaciers, further obscured the state's and mining companies' responsibilities toward them. The central flaw in Conama's argument lies in the agency's arbitrary distinctions between "certain" and "uncertain" scientific information. By calling water loss studies uncertain, Conama could apply the precautionary principle. But by ignoring other truly uncertain issues, the agency could ignore far more serious questions regarding how mining accelerates glacier loss.

A central question at stake in Pascua Lama's second EIA had to do with how much water might be lost downstream if the three glaciers disappeared. Since Barcaza and Wainstein's first report for the activists, scientists had maintained that it was more honest to compensate for the loss of the glaciers than try to preserve them. This assessment recognized that mining activities like explosions, traffic, and heavy machinery negatively impact glaciers, but also that in this case, the three glaciers were already so small that losses to downstream water flows would be negligible. Several studies by EcoNorte and the Water Agency similarly concluded that theses glaciers did not contribute much to downstream water flows (Barrick 2004, 2005a,

2005b, 2006). In the second EIA permit, though, Conama (2006) treated this information as uncertain:

> In a worst-case scenario, assuming the total loss of glacial ice in the area of the mine's open pit, and without implementing any of the proposed compensation measures, the impact on water flow is estimated at 2.5 liters per second on average each year, or nine liters per second in the driest month of February. This measure has been validated by the Water Agency, given that this agency calculated an even lower potential impact of 2.5 liters per second in February. (130)

> Considering the information [Barrick] supplied during the environmental impact evaluation regarding the glaciers; definition and quantification of the impacts of removing these; is considered insufficient, leading to…uncertainty about the impacts on the water flow of the Toro and Estrecho Rivers. Further considering that the central compensation measure is vague in terms of availability of water for a dam, location, management, useful life, and maintenance; [Barrick] will have to access the mineral resources without moving, destroying, or intervening the Toro 1, Toro 2, or Esperanza glaciers in any way. (173)

These two passages point to the first contradiction in Conama's rationale. On the one hand, the agency recognized that the company's and state's estimates both conclude that the three threatened glaciers contribute small amounts of water to the rivers downstream.[31] But on the other hand, Conama considered these results uncertain, along with several other issues that were not studied during the EIA process.

In another contradictory turn, the agency laid out how dust accumulation will diminish a glacier's albedo, which would likely increase the rate at which it thaws. But then Conama (2006, 131) dismissed the importance of scientific uncertainties around industrial impacts on glaciers and permafrost from dust, explosions, and traffic:

> With respect to impacts from particulate matter [dust], the Water Agency's regional office says that the following are unknown: the real effects of industrial dust on the Estrecho and Guanaco glaciers, the presence of ice in permafrost (and its importance as a water resource), and the management of snow moved from the industrial installations accumulated in the area. In all these cases, the company says it will do studies and will answer Conama's questions once the study is approved.

These passages illustrate Conama's willingness to arbitrarily render some information certain and other information uncertain. To be sure, and in Conama's own statement, many things about Pascua Lama's impacts remained uncertain: above all, the feasibility of the proposed dam, but also impacts from dust, how best to "manage" snow at the mine site, and a range of physical processes that affect glacier thawing. For example, dark-colored dust from soot or some soils often accelerates glacier melt. Protecting glaciers therefore requires limiting their exposure to certain types of dust that might be released by industrial activity. Yet in Pascua Lama's permit, Conama (2006, 163) writes, "The glaciers in the area of Pascua Lama currently have large quantities of dust naturally deposited on them." The agency thus drew a tenuous distinction between "industrial" and "natural" dust, and then required Barrick to submit studies about industrial dust at a later date, after the project was under way. In the same permit, Conama was hence simultaneously holding Barrick to a high standard regarding water losses but to a low one regarding the impact of dust on glaciers.

Conama also dismissed accusations that the three glaciers were small as a result of mining activity in the area. As part of its EIA review process, the Water Agency sent its in-house experts to inspect Pascua Lama. The Water Agency's glaciologists concurred that Toro 1, Toro 2, and Esperanza were too small to save, but they pinned responsibility for their small size on mining activities. In a memo that was written to be short "so everyone would read it," the Water Agency concluded, "We can discard the company's theory that the reduction in the glaciers' size is due to climate change."[32] Others corroborated these findings, including a scientist's peer-reviewed journal article that found Barrick negatively impacted Pascua Lama's glaciers (Brenning 2008), a student's thesis redacted with support from several of the glaciologists involved with the case (Bórquez 2007), and glaciologists' private admissions during interviews. One glaciologist raised his hands and crisscrossed his fingers to represent the roads he saw over the three small glaciers. Another described an enormous photograph hanging on the wall in the cafeteria at Pascua Lama's base camp. The photograph showed massive machinery sitting on top of one of the impacted glaciers. Later, while addressing an investigative congressional committee, this scientist was asked whether Barrick had already damaged the glaciers. He assented. The recorder then asked, "You mean, there was not very much scientific rigor in [the consultant's] studies?" and he assented again.[33]

Despite Conama's decision, Barrick and the Irrigation Association executed most of their agreement; the association received US$60 million from Barrick, but the compensatory dam was not built. To opponents, Conama's decision to issue the mine's EIA permit reflected a combination of corruption and incompetence, or the victory of political over technical criteria, as Conama again succumbed to pressure from the executive government, which was as always committed to the project. State officials close to the decision-making process saw things differently, however: in their view, the executive government had made a political decision to protect the glaciers (presumably in response to public pressure) and in parallel (that is, independently) Conama had arrived at a technical rationale to justify that same decision. In this light, Conama's appeal to the precautionary principle was a creative attempt to legitimate the executive government's newfound commitment to protect the glaciers.

Conama's decision-making style, though, weakened the foundations for democratic accountability in at least three ways. First, the result and rationale of Conama's decision occludes any aspect of this case that might be construed as an environmental victory. Taken at face value, several aspects of this decision should have been a cause to celebrate democracy at work: Conama did not toe the line set by Barrick and industrial farmers in the association. Instead, the agency applied the precautionary principle to protect glaciers, making them part of a "national patrimony" in need of special protection. Yet to activists and community members, Conama's decision was anything but a victory. They saw any approval of the mine as evidence of a policy of completed actions because although the three small glaciers had been "saved," Barrick was not held responsible for past, present, or future damages to these glaciers, as the Water Agency (and activists) had called for. For activists, this was a procedural injustice compounded by Barrick's strategy of dividing the community through aggressive social spending that rewarded individuals—especially the association's leadership—who abandoned their opposition to the mine. The net result of Conama's decision was increased distrust and skepticism that Chile's democratic institutions exist to serve common citizens.

Second, while Conama's decision reflected a desire to save the three little glaciers, it missed the bigger picture: the large-scale, accelerated thawing of glaciers due to the compounded impacts of climate change and mining activities.[34] To use the agency's own terms, Conama refused to technically examine a politically important question: Was Barrick responsible for damaging the glaciers, as the Water Agency claimed, or had the glaciers been lost to climate change? If only climate change was responsible for their shrinkage, why go to such lengths to protect the glaciers from mining activity? In denying the range of direct and indirect impacts that mining activity has on nearby glaciers, Conama was embarking on a quixotic quest to protect these three small glaciers. The permit's requirement that Barrick "access the mineral resources without moving, destroying, or intervening the Toro 1, Toro 2, or Esperanza glaciers in any way" was both idealistic and impractical; the glaciers sit on the edge of the mine's open pit. Infrastructure, machinery, debris, and dust fill the area. Conama's decision thus reflects the state's denial of the slow violence facing the Huasco Valley's agricultural communities.

Third, Conama's decision was simultaneously transparent in that anyone could download the EIA documents from the Internet, and opaque in that the foundations for decision-making were ambiguous and contradictory. As a result, subsequent efforts to hold Barrick accountable for damaging glaciers have come to naught. In March 2015, Chile's Environmental Tribunal acquitted Barrick for glacier damages, arguing that Conama set the company up to fail by protecting Toro 1, Toro 2, and Esperanza (*Cruz Pérez v. Barrick* 2015).[35] The judges first admonished Conama for failing to consider Barrick's offer to compensate for the loss of the glaciers. In the judges' view, the scientific consensus at the time recommended compensation, given the results of water loss studies; it was therefore Conama's failure, not Barrick's, that the glaciers had been damaged. The judges then worried about how to compensate losses resulting from climate change. Although their decision absolves Barrick, the court at least recognized that the Huasco Valley's ecological conditions will continue to deteriorate at an accelerated pace due to Barrick's activities there. Unlike other representatives of the Chilean state, the judges recognized that business as usual cannot continue given the realities of climate change.

The judges also lamented the lack of credible science about Pascua Lama. Yet few glaciers in Chile have been as extensively studied as those around Pascua Lama. In addition to the studies conducted for each EIA, Conama required that Barrick set up an ambitious glacier-monitoring plan. Scientists hired to do these studies have generated enormous amounts of data, often using cutting-edge equipment to advance knowledge of basic glaciological processes. Through the monitoring plan, nearly all the glaciologists who worked on the Pascua Lama case, for whichever side, still work on it. Their collective efforts have nonetheless failed to produce something that the judges recognized as credible science, in large measure due to glaciologists' mutual disagreements. For example, although they agreed that the

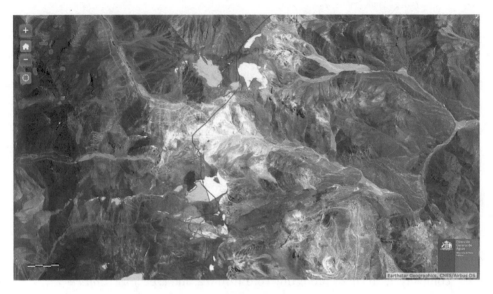

Figure 5.3
The Pascua Lama mine is the large light area in the center of the image,
straddling the Chile-Argentina border (dark gray line). Glaciers on the Chilean side
are outlined and shaded in light gray. Image from the National Inventory
of Glaciers. Created and distributed by Dirección General de Aguas
(the Water Agency) for public use.

monitoring plan did not provide appropriate information for regulatory decisions, the glaciologists disagreed on how to reform it.

Pascua Lama's monitoring data are reviewed by the new Glaciology Unit within the Water Agency, created after the Pascua Lama case to strengthen the state's capacity to review future EIAs from proposed industrial projects that may impact glaciers. Even this new unit has not been able to foster agreement among Chile's small community of glaciologists. The unit's early work, for instance, centered on creating a National Inventory of Glaciers (figure 5.3).[36] While some glaciologists supported accepting industry funding to grow the inventory faster, others feared the inventory would lose credibility as a result. Several years after the unit was up and running, a proposed "national strategy for glacier management" still sat in a drawer waiting for glaciologists to agree on how to validate any shared national strategy.[37] In short, like their patrons and peers in other sectors, Chile's glaciologists disagreed on what makes science credible in the public sphere.

Conclusion

Far from serving as a model for a new generation of high-altitude mines, Pascua Lama has been at the center of public controversy, regulatory scrutiny, and financial skepticism.[38] Analyzed as a case of slow violence, the fight around the Pascua Lama mine illustrates how a normally out-of-sight ecosystem like glaciers can acquire symbolic value in public debates, standing in for multiple, sometimes-competing values. This was achieved in part by glaciologists, who played a variety of roles in the mine's EIA evaluation, with some unexpected twists that shed light on governance in Chile. On the one hand, the Pascua Lama case is evidence of how flawed EIA evaluations can be in Chile. Conama approved the first EIA although it promised to do the impossible (preserve glaciers by moving them), and the second EIA was marred by questionable processes, from the closed-door Barrick-irrigators agreement to Conama's rationale that dodged the most important issues raised by glaciologists and concerned citizens. On the other hand, the case shows how activists can force major changes in the political and legal status of a natural element (glaciers), and demonstrated for the first time that rural communities can challenge mega-mining projects (Li 2017; Urkidi 2010).

Activists' success lay in part in their decision to open up spaces for glaciologists to participate. The first EIA, so lean on scientific arguments or data, offers evidence that Barrick and the Chilean state thought it would be easy to force Huasco Valley residents to transition from agriculture to mining. In this view, Pascua Lama was a welcome project that would bring growth and development. Against this narrative, activists identified glaciers as a potential weak link—an aspect of the EIA that could be easily disrupted in this case by fostering the production of scientific knowledge. This strategy promised to ratify and expand what the community knew about the area, build alliances with concerned actors outside the Huasco Valley, and raise Barrick's costs. Absent this strategy, it is unlikely the second EIA would have been laden with scientific arguments, advanced mitigation and monitoring technologies, and multiple environmental safeguards that were absent from the first EIA permit. Activists could not, of course, control who participated or the trajectory of the debates. Thus, how glaciologists were selected and participated sheds light on the ethical and socioeconomic gulfs that characterized each interest group (activists, the company, irrigators, and the state). Once again, Chileans disagreed on what made for a good, trustworthy scientist.

Another noteworthy aspect of this case relates to the Water Agency's in-house glaciology science capacities, which were expanded post–Pascua Lama to strengthen the state's knowledge of glaciers through an inventory. The inventory promises to capture an objective nature with aerial and satellite images of Chile's glaciers in order to fill a blind spot that Pascua Lama exposed: state officials could not evaluate impacts to glaciers in EIAs if they did not know that glaciers were there. The Glaciology Unit, the inventory, and Pascua Lama's ambitious monitoring plan greatly expanded glaciologists' opportunities for participating in public decisions affecting glacial ecosystems.

But how productive will glaciologists' input really be? Caution is warranted. First, the Pascua Lama case suggests that glaciology is repeating a trend observed in the previous two chapters, whereby the more scientifically studied an area is, the less scientists and officials profess to know about it (they instead emphasize the "unknowns"). Second, the glaciologists have yet to change how the state values glaciers, which continue to be water resources evaluated on a case-by-case basis. The inventory supports this approach; for instance, images are not dated (figure 5.3), making the inventory a snapshot that leaves no public record of changes to glaciers over time. It is an approach permeated by fatalism. The rationale is that given climate change, glaciers are bound to disappear and therefore not worth protecting from profitable mining activities. By contrast, an empowered and more ecologically sound rationale would put glaciers first in line among natural resources needing special protection due to their vulnerability to climate change and importance to stable water supplies (Taillant 2015).

Finally, the case illustrates the coproduction of technical and political criteria, with the terms frequently invoked as foils for each other. As used by the interest groups involved in Pascua Lama, sometimes technical appeared to mean scientific, as when the activists enrolled glaciologists to help legitimate their claims to a broader public. Some studies have warned about the dangers of scientization, when scientific claims obscure value-based claims, marginalizing political considerations from public debates (Kinchy 2012; Li 2015). The Pascua Lama case, however, reveals a more complex pattern. Although the glaciologists did not reciprocate activists by opening up spaces for them to participate in the EIA evaluation, neither did they obstruct citizens' access to state authorities, challenge the legitimacy of public concerns, nor impose an exclusively scientific definition of what constitutes a valid argument. Instead, scientific and nonscientific claims coexisted; if anything, the participation of glaciologists helped Huasco activists gain allies in society at large.

More commonly, though, the term technical invoked ideals of procedural justice that EIAs have failed to live up to. Activists expected Conama and the other agencies involved in Pascua Lama's EIAs to consider all the available knowledge and, with intelligence and public sensitivity, balance competing economic and environmental goals. This was in contrast to what activists called the policy of completed actions: an EIA decision made from the get-go rather than as a result of a balanced evaluation. In particular, activists opposed the economic unfairness of it, as the state privileged mining over the interests of farmers.

Conama staff shared this definition of political as something that comes from the executive government, but used technical in a different way: what made Conama's decision to introduce the precautionary principle technical was that the idea originated within Conama as state officials flexed their capacities and exercised their authority. It did not originate, and indeed had nothing to do with, the logics used by the executive government or members of the public. Yet in imposing sui generis judgments of what was certain and uncertain scientific knowledge, Conama's rationale subverted classic liberal foundations for accountability in EIAs. These should include scientific representations of nature that citizens and investors alike are able as well as willing to accept as evidence for a given decision. Instead, in the Pascua Lama case, Conama's statements prevented the creation of a stable, coherent set of technical claims that might have made transparent the foundations of Conama's decision. The next chapter examines such efforts to improve the technical foundations of EIAs by analyzing a conflict that erupted earlier in the EIA process, before data collection even started.

6   Seeing Like an Umpire State: The HidroAysén Dam Project

On one summer night in January 2011, a crowd joined President Piñera along the Mapocho River in downtown Santiago. They gathered on the river's banks to inaugurate the city's Museum of Light. Twenty-six projectors beamed 104 images along one kilometer of the Mapocho to celebrate Chile's natural landscapes (Weber 2011). As the president opened the show, activists in kayaks maneuvered up the river and unfurled a banner, asking, "Mr. President, will you reject HidroAysén?" (Ambientalistas 2011; Schaeffer and Smits 2015). HidroAysén was a flagship project to build five hydroelectric dams 2,300 kilometers south of the capital in Patagonia's Aysén province. Elites insisted that Chile needed the energy to grow. Activists challenged them, questioning whether Chile really needed more energy given how much of it was wasted, for example, on shows like the Museum of Light. HidroAysén also threatened the landscapes that the museum claimed to be celebrating, particularly the Baker River, an iconic destination for international tourists interested in kayaking. While proponents claimed Hidro-Aysén would provide the energy Chile needs to develop, those against it rejected industrializing Patagonia to benefit industries further north.

That year, the EIA Agency—which replaced Conama in 2010—had to decide whether to approve or reject HidroAysén. As with all such projects, HidroAysén would be built if it received EIA approval. This time, the EIA process involved thirty-six state agencies, municipalities, and public services that would evaluate HidroAysén's baseline and impact studies, and recommend, based on this so-called technical information, whether to approve or reject the project. Against this official narrative, activists denounced corporate and political influences on the decision-making process. The decision did not actually lie with a technical EIA Agency, activists feared, but rather the political authority of the president. Chileans were as divided over HidroAysén as they were over how the decision to approve or reject it should be made; many *Ayseninos* demanded the project's future be decided not in Santiago but instead by authorities in Aysén; others defended the need for a technical decision premised on laws and facts (Mena and Rivera 2011; Martínez 2011); and still others argued that the government needed to decide based on a shared vision of the country's future (Mansuy 2011; Mosciatti 2011; Navarrete 2011).[1]

Five months after the kayak protest, the EIA Agency approved Hidro-Aysén, contending that the project met the requirements of all applicable laws. This gave Endesa, the company that owns the project, a green light to build two large dams on the Baker River and three on the Pascua River, with an installed capacity of 2,750 megawatts (figure 6.1).[2] Crowds gathered outside the EIA Agency's Aysén office to challenge the decision, and thousands more marched in cities across the country. An estimated forty thousand marched in Santiago—the largest demonstration since those held to bring about the end of Pinochet's dictatorship in the late 1980s. Opinion surveys reported that over 60 percent of Chileans rejected HidroAysén.[3]

HidroAysén soon stalled (table 6.1). Opponents objected that Hidro-Aysén's EIA had failed to address several issues: negative impacts to flora and fauna, heightened risks of glacial disasters, deterioration of coastal water quality, and legal requirements regarding fourteen families that would be displaced. Meanwhile, Endesa objected to how the EIA Agency was handling the case; in particular, the agency required an audit of the scientific studies that had gone into the original EIA. This was an expensive effort that amounted to a technical review of what, in theory, had been one of Chile's most thorough EIA studies to date. Endesa lost the fight to legitimate its project, and in June 2014, the Council of Ministers that examines appeals to EIA decisions decided to revoke HidroAysén's EIA permit.[4] By 2017 it was clear that HidroAysén would not be built after all.

Were the decisions to first approve and then reject HidroAysén based on politics, science or technical criteria? What did these labels mean to the different actors in this controversy, and how did these meanings shape perceptions of the HidroAysén decision? In the making of HidroAysén's EIA studies, Endesa and some scientists operated from a liberal ideal of science as inherently independent and detached from human interests. Endesa accordingly hired prestigious scientists from Chile's research universities

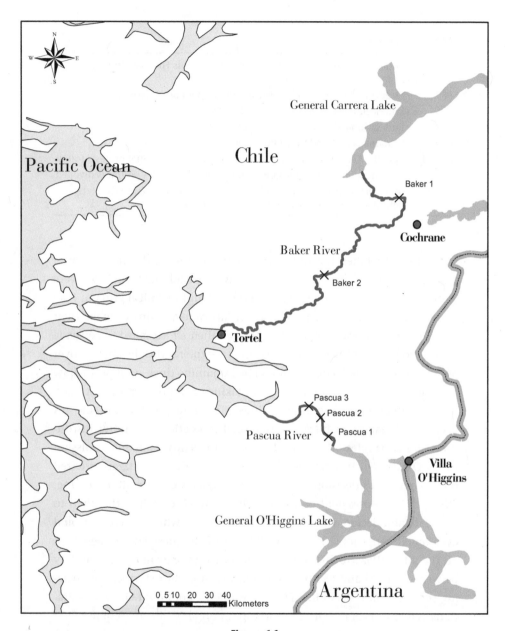

Figure 6.1
Map of the five dams of the HidroAysén project: two on the Baker River
and three on the Pascua River.

Table 6.1
Timeline of Events

1970s	Endesa, a state-owned energy company, designs HidroAysén with aid from Japan
1989	Endesa is privatized, following a series of neoliberal reforms
2005	Endesa announces HidroAysén will be built
2008	Endesa submits HidroAysén's EIA to Conama
2010	Conama is replaced by an EIA Agency
2011	The EIA Agency approves HidroAysén's EIA, and national protests follow
2015	The energy minister announces that HidroAysén's EIA permit is revoked
2017	Investors return the water rights needed for HidroAysén to the state, definitively ending their interest in the project

to do the EIA studies. But they were unable to impose this liberal vision on others, who rejected it in different ways. Instead, public debates were dominated by efforts to distinguish political and technical criteria. Like Conama officials in chapter 5, state officials were at pains to make their work appear technical, not political, though their efforts failed to persuade those who opposed the project. This chapter explores this disjuncture over politics, science, and technical criteria by examining three attempts to draw a boundary around science during the making of HidroAysén's EIA studies.

This case also calls attention to elites' main concern: that unstable EIA decisions threatened Chile's international reputation for strong institutions and clear rules—the qualities that had recently gained it admission to the OECD.[5] For instance, former Endesa CEO Hernán Salazar criticized the government's decision to revoke HidroAysén's EIA permit, observing, "[It is] as if the president of a soccer club wanted to go into the pitch to play ball, and decide who plays and who doesn't, when the coach would do a much better job. The responsibilities of the president are best kept off the pitch" (Pizarro 2014). Salazar thus tried to draw his own boundary between political and technical criteria, which if blurred, risked politicizing such technical decisions. This, he warned, "does not befit our aspirations as members of the OECD"—an aspiration for Chile to become a First World nation (ibid.). Speaking about HidroAysén, President Bachelet also addressed these concerns when she reaffirmed "the principles that should guide our national development: respect for environmental regulations in all sectors of the economy and the presence of stable and clear rules so that all actors, public and private, may contribute to Chile's growth" (Astudillo and Pérez-Cueto 2014).

Technical is an ambiguous term in Chile, used differently by politicians, officials, professionals, and activists, and it captures aspirations for decisions to be "scientific" and "legal" as opposed to political. This chapter explores the negotiations around the technical/political boundary, understood as Chile's version of the science/policy boundary, to reflect on the relationship between rules and science in an emerging democracy influenced by neoliberal principles. The chapter first describes HidroAysén, a classic example of high modernist development, and the challenges it posed for Chile's government after neoliberal reforms were introduced. The second half focuses on three forms of boundary work practiced during the making of HidroAysén's EIA. The final two sections discuss the effects science had (and failed to have) on the state's decision to approve the project.

A High Modernist Project

HidroAysén consisted of five large dams to be built on two rivers in Aysén, a region with little infrastructure. Construction would have required relocating five thousand workers to the region, thereby doubling the local population. These new residents would have lived in four purpose-built workers' camps, with new waste and wastewater treatment facilities as well as new medical facilities. To build HidroAysén, Aysén's roads also would have needed major upgrades. Only one road—the misnamed Carretera Austral, or southern highway—traverses Aysén from north to south (figure 6.2). Most of the road is unpaved and at times narrow; drivers have to move aside to let trucks that deliver fresh fruits and vegetables pass. The tiny port of Yungay, which has facilities to receive a barge large enough for twelve cars to cross a rainy fjord, would have been transformed to accommodate a ferry large enough for heavy machinery and tons of cement. HidroAysén's social impacts would have been most acutely felt in Cochrane, a town of three thousand, and its environs. The nearby Baker I dam and its reservoir would have inundated parts of the Carretera Austral and displaced fourteen families. After years of protests, the government finally recognized that it would be impossible to relocate these families as a single community to comparable ranch land.

The enormous gap between Aysén's rurality and the industrial intervention HidroAysén required fueled much of the opposition. Many believed

Figure 6.2
The Carretera Austral (the southern highway) south of Cerro Castillo.
Photograph by author.

Aysén would gain little from the project, as the electricity would be sent north rather than used to lower the high electricity rates Ayseninos pay. Environmentalist and scholar Nicolo Gligo (2011) explained opposition to the dams as frustration with distributive injustices: "In the pursuit of likely economic growth that they know will reach them little if at all, those opposed to HidroAysén do not want to sacrifice part of their territories' natural identity." The satirical weekly the *Clinic* (issue May 12, 2011) mocked the kind of development HidroAysén would bring—small apartments, dirty public transit, boring office work, and high personal debt—suggesting that Chile's modern cities were no example for Patagonia to follow. An NGO, Patagonia without Dams, coordinated different opposition groups and launched some effective campaigns. Within Aysén, these campaigns appealed to themes related to preserving Patagonian traditions and lifestyles threatened by HidroAysén. Nationally, the campaigns appealed to a shared sense of awe in the beauty of Patagonia's landscapes (Schaeffer and

Smits 2015). These concerns resonated with Chileans, as illustrated by these quotes from the public consultation for HidroAysén's EIA:

> What does Aysén really gain in the long term from this project? (EIA Agency 2011, 696)

> As a citizen of Chile and Aysén, I don't want this project to be built, I have no desire to see any impacts in one of the few places on earth where there still is a wild ecosystem. Enough of distorting sustainable development. HidroAysén will finish off one of the most pure identities of our nation. (Ibid., 677)

Anti-HidroAysén discourse also reflected an antigovernment sentiment. Political strategist Jorge Navarrete (2011) explained that opposition to the dams demonstrated that "[Chileans] no longer want to be treated as beneficiary, consumer, or client, but as a partner." Citizens, Navarrete wrote, had tired of a "paternalistic and neoliberal," state and demanded to be treated "not as subordinates but as equals." For Ayseninos, being treated as equals meant challenging Chile's notoriously centralized government, in which regional governors are appointed by the president and state officials answer to superiors in Santiago. It also meant highlighting what they saw as Santiago's neglect of Aysén, evidenced by the region's poor infrastructure. For decades the state had failed to bring development to Aysén, and to pretend to do it now with HidroAysén was an injustice. The project contravened the regional government's sustainable development agenda, based on tourism, specialized agriculture, and aquaculture, without proposing a coherent alternative. Instead, HidroAysén would be the first of many dam projects in Patagonia to disrupt local lives and benefit a small number of corporations.

The dream of HidroAysén originated in the postwar era, when states pursued public works—particularly dams—to deliver modernity and development (Khagram 2004). The scientific and technical advances made during World War II created new possibilities to improve quality of life. Political leaders who could harness these advances could then promise to maximize agricultural productivity, generate cheap electricity, and improve citizens' welfare. Postcolonial states worldwide turned to science, technology, and central planning to consolidate state bureaucracies as well as bolster new leaders' popularity. This was the high modernist turn, as states built up their scientific and technical capacities, used these to articulate coherent development programs, and then unleashed these on populations using an "imperial rhetoric" of civilizing peasant communities (Scott 1998).

Throughout the twentieth century, Aysén was the target of such interventions from the Chilean state. Late in the nineteenth century, descen-
dants of Europeans began to arrive in Aysén in search of land for ranching
and sheep herding. Pioneers displaced indigenous peoples and torched four
million hectares of forests to create pastures. After 1900, the Chilean state
began to intervene, formalizing land grants along with introducing measures to coax settlers and their goods to Chilean ports, away from more
accessible Argentinean markets (Ivanoff 2007). Access to Chilean markets
grew slowly, but by midcentury the Chilean state was interested in the
region's hydroelectric potential.

Two enormous ice fields and countless other glaciers dominate Aysén.
The region also receives plenty of rain and snowfall. The snow and ice
melt into South America's two second-largest lakes (after Titicaca), each one
named after two of Chile's founders, General José Miguel Carrera and General Bernardo O'Higgins. From these lakes, rivers rush down the mountains,
reaching the coast in less than two hundred kilometers. As a result, Aysén's
rivers are powerful. When HidroAysén was first proposed in the 1950s, energy
was the responsibility of Endesa, then a state-owned company that embodied the nation's engineering expertise (Tironi and Barandiarán 2014). In the
1970s, with assistance from JICA, Japanese and Endesa engineers collected
data along the Baker and Pascua Rivers, installed water-monitoring stations,
and designed the original HidroAysén project.

After 1973, HidroAysén remained central to the military government's
designs on Aysén. Patagonia has geopolitical significance because the border there with Argentina has often been contested and remains porous.
The rivalry with Argentina made developing Aysén a priority for Pinochet's
military government, and roads and electricity were the way to do that.
Villages like Caleta Tortel (population 320) received small hydroelectric stations to cover their needs for free, while HidroAysén would provide cheap
electricity at regional and national scales. But the costs of HidroAysén, coupled with the privatization of Endesa, meant the dam was not built.

The military government, moreover, ran into difficulties when building
the Carretera Austral, a highway that was supposed to have modernized the
remote region (figure 6.2). Using forced military labor, the government set
out to build a highway that ran Chile's entire length. Young soldiers tore
through temperate rain forests, leveled hillsides, and crossed rivers and
streams on their march south. Dozens died. But the road remained unfinished

into the twenty-first century. The bus I traveled on took eight hours in good weather to cover three hundred kilometers between Aysén's capital city, Coyhaique (population 58,000 in 2011), and the next-largest town, Cochrane.

High Modernity Meets Neoliberalism

HidroAysén's contemporary iteration differs in critical ways from its 1970s' predecessor. The project's new design reduced the area to be inundated to mitigate some of the dam's environmental and social impacts. More important, the political context changed dramatically between the 1970s and 2000s. The first major change involved water rights and electricity regulation. In the early 1980s, Pinochet's government introduced legislation that separated water from property rights and created a market for water users (Bauer 1998). The government also deregulated electricity generation; power companies only need EIA approval, rather than a concession, to sell electricity to large users, or on a regulated market for small and residential users. Together, these reforms have favored hydroelectric power (Prieto and Bauer 2012).

Another major change involves the privatization of Endesa during the last gasps of the military government. The sale included water rights for hydroelectricity on the Baker and Pascua Rivers as well as the design for HidroAysén, along with supporting stream flow data and monitoring stations. Prior to privatization, Endesa functioned as a de facto energy ministry. Instead of replacing the state-owned Endesa with a ministry, following neoliberal principles, Chilean policy makers shifted their attention to collecting data on Chile's electricity market while leaving decision-making in the hands of private corporations. Thus, the twenty-first-century version of HidroAysén belonged to a multinational controlled by the Italian government's Enel company (this firm is also called Endesa, and owns 51 percent of the project), and a Chilean family business group, Colbún (49 percent). To supporters of privatization, Endesa's sale was intended precisely to get the government and other such political voices out of energy policy (Prieto and Bauer 2012; Tironi and Barandiarán 2014). Opponents protested that corporations pursue only profits, not the national interest.

Scientific research also changed dramatically in this period. As discussed in chapter 2, the military government reorganized universities and reduced research funding. From an already-low level, the number of active scientists in Chile fell during the 1980s (Allende et al. 2005). The type of science produced shifted too. Before the reforms, most research was state funded and published in broad-ranging textbooks like *The Bioclimate of Chile*. State-funded research at the time was primarily undertaken to foster industrialization. For example, Corfo published a 1993 study of Aysén's climate to develop the region's wind energy. But as the organization of science changed, so did its results: research became more specific and was published in shorter journal articles (Barandiarán 2015a). This published research, however, is rarely used in EIA baseline and impact studies, which instead draws on consultants' reports that are not public. Aysén was especially "unknown" to scientists. Before 2005, no scientists lived or worked in Aysén full time. State-owned monitoring infrastructure was scarce; what existed had been installed by Endesa and privatized with the company. Many scientists who worked on HidroAysén's EIA studies in the late 2000s began recollecting their experiences with a warning: "HidroAysén pushed against the limits of the knowable."[6]

Given this context and anticipating the controversy that was to follow, Endesa turned to what it felt was the best-available science to legitimate HidroAysén. The company hired teams led by some of Chile's most prestigious biologists, ecologists, zoologists, and hydrologists from every research university in the country. Scientists who participated saw in HidroAysén an unprecedented opportunity to explore normally inaccessible territories and work in large teams; the biology team alone included thirty scientists, who flew around Aysén by helicopter and had access to 4x4 Jeeps, studying flora, fauna, insects, and amphibians in the dams' area of influence. A scientist estimated that Endesa had spent the equivalent of eight to ten years of scientific research funded through regular government sources on HidroAysén's EIA.[7] Endesa spent US$12 million on the EIA studies, which amounts to about 14 percent of the national budget for research projects in 2008, the year the studies finished.[8] The intention, according to an Endesa engineer with responsibility over the EIA studies, was to signal the company's commitment to "doing things well," without cutting corners.[9] Science offered Endesa a way to expand society's knowledge frontier through studies of little-known things, like Patagonian lichens and moths. It demonstrated the company's commitment to a transparent and fair process: better science meant a better EIA evaluation. This strategy, the engineer hoped, would avoid the mistakes made by Celco Arauco's Valdivia mill and Pascua Lama's EIAs: HidroAysén's EIA would be judged for its scientific merits, not its politics.

Scientists spent three years working on HidroAysén's EIA studies. The EIA promised to be epic. In all the promotional material it produced, Endesa touted the EIA's statistics: more money had been spent on HidroAysén's EIA studies than ever before in Chile's history, relying on the best scientists from eight universities (HidroAysén 2008). Inside state agencies, officials received these statements with some anxiety because on receiving the EIA reports, they would have just 120 days to review all the materials and decide whether to approve or reject the project.

On receiving and reviewing the EIA studies, each one of the thirty-six agencies had to send their questions and issues back to the EIA Agency, which then compiled the responses and submitted them to Endesa for a response. Endesa then stopped the 120-day clock to elaborate its answers, submitting them back to the EIA Agency, which in turn distributed the replies back to the relevant agencies. On receiving Endesa's responses, the EIA Agency resumed the 120-day clock where Endesa had stopped it. This revise-and-resubmit process continued until the state agencies' 120-day clock ran out; at that point, the EIA Agency had to make a decision based on what it had. The longer the EIA, the harder it would be for the understaffed state agencies to meet their goals within this time limit.

The day Endesa submitted HidroAysén's EIA, trucks arrived early at the EIA Agency's office in Coyhaique to deliver boxes full of documents; HidroAysén's EIA was a "cubic meter of information."[10] The cubic meter was distributed to offices in Aysén and Santiago, as offices in the capital lent support to their Patagonian colleagues. HidroAysén's EIA was a major effort for state agencies: staffers were reassigned to work on this project, desks had to be cleared to make room for the documents, and new computer programs were installed so staff members could view the EIA documents. After all this preparatory work in anticipation of HidroAysén's scientific, high-quality EIA, it came as a surprise to find that HidroAysén's EIA maps were illegible. The pdf documents Endesa submitted for evaluation contained grainy maps that were too small to read.[11]

High modernist states, like transoceanic empires before them, used maps as tools of empire. European empires typically sent mapmakers along with military expeditions to map the territory they hoped to conquer (Burnett 2000), as did newly independent states like Tanzania (Scott 1998), Argentina and Brazil (Andermann 2007), Colombia (Appelbaum 2013), and Chile (González Leiva 2007). In all these cases, mapmaking was a strategic concern of the state. Under neoliberalism, by contrast, mapmaking has become a privatized formality—another box to check on a long list of requirements. At Endesa, a staff person explained the illegible maps by saying, "We thought everything had to be in the same format."[12] Endesa's scientists and the consultants who prepared the EIA documents evaded responsibility by blaming others or their own lack of experience.[13]

Whatever the reason, these illegible maps posed many problems for the EIA's credibility. State officials lost days off their 120-day clock while they waited to receive new, legible maps and software with which to view them. Citizens who decided to participate in the public consultation complained it was a sham; one angry resident stated, "The graphic description of the drawings and maps needs to be improved because they are illegible. [The graphs and maps] have no legend, and no references to areas or places to identify where along the Baker or Pascua River they are referring to, for example. If the intent is to do a public participation process, at a minimum there should be an effort to make the graphic material easily readable and legible to the people" (EIA Agency 2011, 777). To top it off, the EIA documents were so big that it was difficult, if not impossible, to download them from the EIA Agency's website. Once downloaded, the pdf documents were not searchable and frequently caused a regular computer to overheat. Despite the EIA Agency's vaunted transparency, these technological difficulties meant that only citizens with a good Internet connection and steely perseverance could actually access the documents.

This episode illustrates the shift in the state's role from that of an empire, with scientific and technical expertise at its command, to that of an umpire, with narrower responsibilities and capacities. Far from having strategic value, in the contemporary umpire state the maps (and EIA studies they were a part of) were merely perfunctory. The umpire state is hamstrung not just by the poor-quality materials it receives from others but also its limited abilities to produce an official scientific narrative. Chilean lawmakers have advocated for indexing EIA studies into a database so as to transform them into an official resource for state officials to use when evaluating future EIAs, much like scientists use peer-reviewed journal articles.[14] This vision, though, assumes that EIA studies have credibility with the Chilean public— an assumption belied by HidroAysén's EIA. This case stands out because in many ways, Endesa invested in science: the effort was well funded and based on the collaborative efforts of dozens of local scientists. Nevertheless, activists dismissed HidroAysén's EIA as misleading and untrustworthy.

Some citizens also challenged the EIA's science during the public consultation process. For instance,

> The documents mention a specific report about the Guaitecas Cypress tree. Despite many attempts to find this report, this was absolutely impossible.... It literally says "the Guaitecas cypress is not a very abundant tree in the area of study...." However, from figure 5.24 of the same document (Distribution of forest types, page 95), it is clear to anyone that forests of Guaitecas cypress are by far the most important in the entire basin.... Clearly, there is an attempt to "lower the profile" of the amount of cypress that will be impacted within the areas of impact. (EIA Agency 2011, 1233)

> There is no attempt to verify whether the GPS coastal data contains errors or not. The best I have seen in geo-referenced data is 20 meters [sic]. I was not able to identify in any part of the report any attempt to validate the information or correct errors. (Ibid., 1415)

> Amphibians: The reports indicate a low diversity and abundance due to the cold climate (sampling was done in May and August with snow and cold), and in addition they measured in the incorrect places. Seventeen species are present in the region, of which 13 are threatened and one is in danger [of extinction]. They found four species: Bufo variegatus, Butrachyla aff. Nibaldo (endemic, low population density), Eupsophus aff. colcaratus (endemic), Pleurodema bufoni (insufficiently known). The report says and recognizes that there must be more species than they found in the area of study. Both the results and their declarations in the study are evidence that the quality of the results is irregular, inconsistent and frankly inclined to error. (Ibid., 1452)

Citizens who participated in the public consultation thus rejected Endesa's claims to have produced a scientifically excellent EIA. They considered the EIA science suspect for reasons that range from concerns about conflicts of interest—as one frustrated activist exclaimed, "In Chile there is no such thing as an independent expert" because none are trusted to be independent of their funders—to doubts that Chilean scientists are really experts on the topics they are hired to study.[15] Chilean scientists who participated in HidroAysén's EIA believed that Endesa hired them because "only [scientists] know how to do this kind of complex study well."[16] But few outside the scientific community agreed with this assessment. Citizen comments to the EIA suggest that they felt they knew Aysén's nature just as well or better than the scientists.[17] Likewise, environmental consultants dismissed university scientists because they lack access to the latest technologies and are not up on the latest research methods, which are available to better-funded consultants.[18] And many scientists dismissed their colleagues who work on EIAs as opportunistic, claiming they follow the money without regard for the consequences of their work. An experienced engineer lamented that "in Chile, the most biodiversity always exists in the most recently studied river, which is always the next one slated for dam construction."[19] The next sections examine how, faced with such criticisms, Chilean scientists tried to defend their credibility.

Boundary Work 1: Science as Baselines

Inaccuracies aside, many scientists knew that the credibility of EIA science lay in the cultural dynamics of science—what STS scholars capture through boundary work (Gieryn 1999). Chilean scientists expressed this with an often-repeated adage: "The wife of Caesar must not only be chaste and pure but also appear to be so."[20] In the marriage of convenience between science and policy, impartiality and independence are not accepted on faith; they must be convincingly performed through boundary work: a set of tactics used to draw boundaries between scientific and nonscientific forms of knowledge (Bijker, Bal, and Hendriks 2009; Hilgartner 2000; Jasanoff 1990, 2005; Keller 2009).

Endesa hired scientists to work on HidroAysén's EIA seeking credibility. A company engineer who oversaw the EIA studies acknowledged that academic scientists brought an appearance of independence that in-house experts lacked. Committed to the company and project, Endesa's engineers were likely to share certain epistemic and cultural perspectives that would favorably dispose them to the project. Furthermore, they were likely to want to speed things along rather than take the time to study things carefully. Better to hire outside experts, uncommitted to HidroAysén, who would pursue accurate studies versus cut corners to have the project approved quickly.

Normally companies trust the entire EIA process to environmental consultants who specialize in preparing EIAs. Unless something specific is at stake—as was the case with glaciers, in chapter 5—companies do not look outside the specialized world of consultants for scientific advice. Moreover, consultants do not typically like working with academic scientists. Not only are their methods outdated, but they also "do not think strategically" and "complicate EIAs." Thus what to Endesa represented prestige and legitimacy, to environmental consultants was absurd: scientists who worked on HidroAysén's EIA did "ridiculous studies" of lichens and moths, which in the hands of regulators, "opened up too many windows" through which to raise doubts and questions about the project.[21]

Scientists were equally critical of companies and environmental consultants. A young scientist who had a second job as an environmental consultant described the relationship between scientists, consultants, and the company as "toxic" because the company has power over scientists: "The thing with the EIA system is that the consulting firms are in a vicious circle because the company pays you to do a study to evaluate the company's project's environmental impacts. The company is judge and jury in its own cause."[22] The concern that scientists lack sufficient autonomy from companies to work objectively was widespread. In this view, financial dependence on the company makes EIA science, by definition, partial.[23] Like the exasperated activist who claimed there was no such thing as independent science in Chile, scientists too were concerned with how to prove that they remained "chaste and pure" despite having entered a marriage of convenience with companies. The solution, many scientists thought, was to focus their contribution on a narrow slice of EIAs: baselines.

Baseline studies are an inventory of the environmental and social conditions present in an area prior to it being transformed by a new industrial project. According to the young scientist and part-time consultant, "Baselines produce technical information.... [T]here is no politics to it."[24] Baseline studies exist in contrast to impact evaluations. Nearly every scientist I spoke with who participated in HidroAysén's EIA described their work as limited only to the baseline studies, which are scientific, as opposed to impact evaluations, which are messy, political, and involve "questions of values." One scientist even claimed that baselines are "so technical, we did not even know where the dams would go when we collected data from the Baker River."[25] Staff at Endesa agreed; the engineer quoted earlier explained, "To prepare baselines is a more pure scientific study, like taking a photo, to see what there is. To evaluate environmental impacts is different." Baselines are like a "photograph" over which "there isn't much to discuss." In his view, "it is very appropriate for universities to characterize the current environment." Impact evaluations instead are "delicate" and require capacities only consulting companies have.[26]

Distinguishing between baselines and impact evaluations had several advantages for scientists and Endesa. First, it reestablished a firm boundary between scientists and consultants, leaving consultants in charge of the impact evaluations—the strategic part of the EIA that is most important to getting the project approved. Second, scientists were able to participate in EIAs while distancing themselves from the results of their work. By claiming to focus only on this narrow, scientific slice of an otherwise-political process, scientists achieved moral distance from the downstream uses of their work. For example, many scientists complained of "cut and paste," a seemingly widespread practice whereby baseline studies are misrepresented in the final, official EIA documents.[27] Several scientists who worked on HidroAysén's EIA expressed sadness and disappointment on reading the final EIA documents because, in their view, some of their best work on impacts had been left out. The allegedly adulterated topics ranged from coastal impacts to representations of fauna to geologic risks. One scientist went so far as to say it "broke her heart" to read how poorly some of the risks her team had detected were represented in the EIA.[28] Despite their concerns, however, these scientists felt their responsibilities to science, the EIA, the company, and the public had finished when they turned in their baseline studies.[29] The baseline-impact evaluation distinction allowed them to feel comfortable with their role in the EIA.

For Endesa, the distinction had another benefit: the creation of an intellectual capital they controlled. HidroAysén's baseline studies occupied multiple binders arranged on two shelving units in HidroAysén's open houses in Coyhaique and Cochrane, set up to provide the public with information about the project. These reports are not accessible through the EIA Agency's online portal (nor in offices in Santiago). Other acts compound this act of privacy: EIA studies do not typically disclose the raw data used in analysis. For instance, regarding water flow averages for the Baker River, the EIA documents report only weekly averages, and cite an unpublished report by company engineers that presumably contains daily or hourly data (Endesa 2008, Anexo D, Apéndice 4, 61). Surprisingly, many scientists and industry staff who worked on the EIA repeated again and again that the data were "made public through the EIA."[30] Even state officials were unclear as to what was public and what was private, however, as evidenced by this response from the EIA Agency to a citizen's inquiry made during the public consultation:

> The information about meteorological stations used in the EIA are found in publicly available publications, available in offices of different Ministries, Research

> Centers and Institutes of Higher Education. Furthermore, this information is protected by intellectual property rights, so the company is not allowed to distribute it on paper or digitally. They can only make references to it. (EIA Agency 2011, 1364)

Blurring what was public and private information, Endesa pushed the boundaries of transparency in its own favor without facing pushback from scientists or the state. This only heightened public skepticism—a problem that the second type of boundary work, discussed below, expands on.

A few scientists believed they could better distinguish science from consulting work by following some simple practices. One consulting agency that wanted to distinguish its work as scientific published its name on all its work, and simultaneously distributed its reports to the client and regulators, so as to make it impossible for consultants or the company to adulterate its report.[31] Another strategy involved "better contracts," though it remains unclear how well this worked. Staffers at Endesa were confident that "we believe, and we have seen, how scientific independence grows each day and is not up for sale.... You can just ask the scientists and they... can be quite critical, which is very valid for the process. I don't need to hire people who agree with the project, I need to hire the right people for certain studies."[32] But scientists often disagreed and alluded to the contract with Endesa to limit their responsibility, as illustrated in this excerpt from a geology baseline report:

> This study was based on the recompilation, revision and critical analysis of existing geological information, principally that available in the Geology Department of the University of Chile. Furthermore, and following the terms of reference [the contract], special attention was paid to the results of the geology study developed by AURUM INGEROC Consultants Ltd....It is important to note that at the time of writing this report, only reports on compiled and photo-interpreted preliminary geology of the Baker River, XI Region of Aysén, and of the Pascua River, XI Region of Aysén, were available. According to [Aurum Ingeroc Consultants]..., this has the character of a preliminary geological map due to the form in which the data was generated and the lack of validation through fieldwork. (University of Chile 2008, 5)

The statement says that the contract with Endesa led scientists to focus on existing data, especially those provided to them by consultants, without validating them in the field. Read as a caveat, the statement limits scientists' scope and responsibility for the quality of the results.

The baseline-impact evaluation boundary speaks to a classic trade-off between relevance and credibility that scientists face when participating in policy. As scientists become more relevant to matters of collective concern, they compromise part of their scientific credibility by sacrificing typical scientific practices, such as peer review. In US or Dutch agencies, scientists deal with this trade-off by negotiating their credibility with different stakeholders—negotiations that occur within institutionalized advisory boards as well as in meeting spaces where regulators are prominent (Bijker, Bal, and Hendriks 2009; Jasanoff 1990; Keller 2009). In Chile, by contrast, this negotiation happens between scientists and industry in ad hoc spaces that exist outside state institutions. Potential solutions Chilean scientists have identified, such as signing one's work or clearer contracts, would help formalize the relationship between scientists and the company, but continue to leave the state out of it. Chilean state agencies thus have not developed the capacities needed to assess and communicate to a skeptical public the credibility of different knowledge producers. Meanwhile, scientists and their patrons in industry narrowed EIA science down to essential and seemingly objective tasks: to metaphorically photograph a landscape and inventory nature.

Neither the process nor the results satisfied many citizens' needs for credibility, nor those of many scientists who spared no derogatory words for their colleagues who limited their work to baselines. Scientists called baselines a *cahuín*, Chilean slang for a "sham" or "charade"; a "random pile of data"; not fit to publish because "baselines produce data, but not information"; or simply, "baselines are not science." Another said it was "stupid" to believe you could produce "a good baseline without knowing where the dams would go." HidroAysén's EIA was one of many projects that had driven some scientists to say "never again" to the EIA and criticize colleagues who "washed their hands" of the poor quality of EIAs.[33]

Boundary Work 2: Public Science as a Private Interest

The case of HidroAysén begged the question, Which responsibilities for generating knowledge about the region belonged to Endesa, and which belonged to the state? An Endesa employee who worked tirelessly to promote the project in Aysén reflected on some of the difficulties the company faced regarding data. Because scientists and officials felt that little was known about Aysén, the company received constant demands for more data from state agencies—data on glaciers, wind speeds, snowfalls, temperatures, sediments, predictions on climate change, and so on. He felt the company had no good way to refuse such requests; the company was vulnerable to never-ending demands for data from state agencies that lacked the resources and capacity to collect these data themselves.[34]

In 2009, while Conama was evaluating HidroAysén's EIA, Congress debated how to reform the EIA. Some legislators advocated for science to play a greater role in EIA decisions. A young legislator evoked the need for "pure science [to be] at the service of the nation" as an antidote to the "impressionistic criteria typical of Chileans" (LH 20.417, 1956). One answer to these problems was to hire a local, state-funded science research institute capable of producing data credible to state agencies and useful for the company. Such a group was to be found in the Center for Ecology Research in Patagonia (CIEP).

CIEP opened in 2005 with funding from the central government. It was one of several institutes across the country created to decentralize scientific capacity and foster science-based economic activity.[35] As HidroAysén was getting started, CIEP was under construction. Between 2005 and 2010, CIEP hired its first dozen scientists, built modern labs outside Coyhaique, and was constructing a small coastal research station in Caleta Tortel, at the mouth of the Baker River. Research related to tourism and aquaculture grew fast, reflecting Aysén's most important economic activities, but teams working on coastal fisheries, forest ecology, glaciers, and snowfall, among other areas, were also growing. In its first years, CIEP installed new monitoring equipment and expanded the region's research frontier. With CIEP, Aysén finally gained a permanent scientific presence funded by the state. Yet as it was primarily staffed with scientists who had recently arrived from other regions in Chile and abroad—CIEP's scientists came from France, Japan, Argentina, the United States, and elsewhere—its scientists were generally humble about being local. Seeking credibility, Endesa thought CIEP was ideal for conducting HidroAysén's baseline studies. CIEP was scientific, public, and in need of building its local reputation as well as experience. A partnership with Endesa seemed like a win-win situation.

CIEP, however, was soon forced to pull out of working on HidroAysén's baselines, with its participation engulfed in controversy and speculation. The official narrative offered by CIEP's second director to explain these events is that because Aysén's regional governor chaired both CIEP and the committee that would later vote to approve or reject HidroAysén, if CIEP had participated in HidroAysén's studies, it was entering into a conflict of interest. The assumption was that the governor would have had an incentive (and ability) to intervene in CIEP's science to favor the project. CIEP's frustrated director noted that for this reason, "[CIEP] could not be seen as a technical, scientific group."[36] Even after CIEP officially pulled out, rumors continued. CIEP organized a workshop to present HidroAysén's (2008) baseline studies as an opportunity to learn about the results of this massive data collection effort. But CIEP was strongly criticized in the local newspaper by a leading activist, who spoke for many skeptical Ayseninos when he said,

> CIEP was created to be at the service of scientific knowledge in our region, and today we see how this original idea is being corrupted as CIEP is giving its support and image to research for a project that still has not been subject to environmental evaluation....It is not healthy for part of the State's resources to go to promotional activities for projects that the State itself has to evaluate. (Es delicado 2008)

Activists contended that CIEP was helping to market HidroAysén by lending the project CIEP's image and worried that state funds were going to promote a private interest.

CIEP's director at the time replied to these criticisms using familiar rhetoric: he affirmed the independence, credibility, and authority of science, as if these were inherent to it. While the activist's critique blurred the boundary between public and private interests, CIEP's director tried to keep science and politics separate. After underscoring that the baseline studies were based on "validated scientific methods" applied by "excellent and virtuous scientists" who work at the country's universities, he added,

> [CIEP Director Eduardo Vera] also stated that to suppose spurious partnerships between CIEP and any company goes against the prestige, nobility, and integrity of the institutions and scientists who are part of CIEP....What the interested community should observe is what representatives from CIEP present and subscribe to, so as to comment on what companies and the authorities should resolve to do or not do based on [CIEP's] research; thus judging in this way those who make decisions based on scientific data, but not politically admonishing those who have scientifically obtained that data. (Director del CIEP 2008)

Although Vera's argument reflects a common ideal of science as independent (Brown 2009), it did not find a receptive audience. Instead, public skepticism was so strong that two years after this exchange, CIEP was still the subject of rumors across Coyhaique. The word on the street among Ayseninos was that CIEP was a consulting agency desperate for funding, competed with universities for funding, or was a puppet of its two founding universities, Austral and Concepción. Because scientists from these universities had participated in HidroAysén's EIA, many felt that it was as if CIEP had participated, thereby preserving the conflict of interest many were concerned about. Like with CENMA (chapter 2), these rumors reflected anxiety around what CIEP is—a public research institute or private consulting group—and who it works for.

At the time, CIEP was funded by several sources: a quarter of its budget came from the regional government, and the rest was roughly equally distributed between the national science agency, universities of Austral and Concepción, two foreign universities, and a few companies that sponsored aquaculture research. Like the University of California and many research institutes around the world, CIEP has a board that includes business and political elites. In Chile, distrust of universities is perhaps particularly strong; one (foreign) consultant said bluntly, "Chilean universities have no prestige to protect."[37] The case of CIEP illustrates the difficulties of constructing new science in this context as well as the pervasiveness of suspicions that state and state-funded institutions pursue private interests rather than act as guardians of the public interest. Moreover, because the regional governor never called on CIEP for advice, the institute never had an opportunity to build a public reputation. As with CENMA, this episode illustrates that building science that might be "at the service of the nation" requires not just new labs but also opportunities for scientists to demonstrate their public credentials.

Boundary Work 3: Science Misunderstood

By contrast to the above examples, which involved scientific organizations participating in EIAs through formal channels, this section explores the experience of two individual forest ecologists who challenged Hidro-Aysén's EIA in more informal ways. Alex Fajardo and Frida Piper, a husband-and-wife team employed at CIEP, became concerned about the quality of two reforestation plans submitted by Endesa and another company, Xstrata, to compensate forest losses from two hydroelectric projects (HidroAysén and Rio Cuervo, then also under EIA review). These plans proposed to "reforest" twelve thousand hectares—an area about the size of the city of San Francisco. Never before had such an ambitious reforestation project been undertaken in Chile, and there were reasons to hesitate. Piper and Fajardo found that these plans were based on protocols from warmer climates, relied heavily on pesticides, and targeted steppes that had never had forests on them. While Endesa promoted it as a reforestation plan, the ecologists called it "afforestation" because it created forests where none had ever existed. Part of the difficulty, Piper and Fajardo explained, was scientists' lack of knowledge about Patagonian forests. The plans proposed to apply practices developed for fast-growing species in warmer climates in Patagonia, where trees grow slowly in cold, sunless weather. "If the reforestation plan contains so many errors and is fundamentally a lie, how can we trust the rest of the project?" they wondered.[38] Concerned, they decided to act.

Speaking up came with symbolic and material costs because many believe that in criticizing a project, scientists are acting politically as opposed to objectively or neutrally.[39] To minimize symbolic losses, Piper and Fajardo pursued their critiques within well-recognized scientific practices. They published their concerns in academic journals, like *Science* (July 23, 2010, 384) and *Frontiers in Ecology* (March 2011, volume 9, issue 2). They obtained the blessing of CIEP's director, who hand delivered a memo with their concerns to the regional governor. There was little they could to do minimize economic losses in the form of lost research funding. But this did not worry them because, Piper and Fajardo noted, their research is low cost. They work through observation, not expensive experiments, in the same forests where they live. Asked why they spoke against HidroAysén when so many of their peers remained silent, they mentioned their childhoods. Piper is from the Argentinean pampas and Fajardo is from Chile's rural south. They grew up surrounded by increasing ecological harms from industrial development and learned to be skeptical of top-down development projects. They did not want to see those patterns repeated in their new home, Aysén.

Endesa tried to meet this scientific critique with what the company called science. A young engineer at HidroAysén's open house in Coyhaique recalled that he first heard about Piper and Fajardo's criticism on the local radio. Back at Endesa, he and some colleagues decided to "rise to the challenge" to address this "defeatist" perspective: "just because something has never been attempted does not mean it will fail." To think like this "is to oppose progress."[40] Endesa had recently done exploratory work on a thirty-acre plot by the Pascua River, where three dams would go, that needed to be reforested following Chilean regulations. It reforested with native trees, and one year later—following regulations—the Forest Agency certified that over 90 percent of the trees were "reforested and growing." Pleased, Endesa's staff members published the results in what they described as "academic journals." A two-page article appeared in three industry magazines: *Lignum*

(January 26, 2011), *Buscagro* (n.d.), and Uruguay's forestry association newsletter (February 1, 2011). No aspect of what Endesa's engineers called a "scientific experiment" met current scientific practices: there was no experimental method, the journals were not peer reviewed, and the articles did not describe what they did or who worked on the experiment. A wide chasm thus separated companies' and scientists' understandings of science.

Furthermore, in this case, what the company intended as signals of credibility, others regarded as the exact opposite. Endesa's engineers described the Pascua reforestation as executed by "the best forest nursery in the country" and "best forest engineer of the region." They explained that the article did not specify who was involved because "the company is responsible for the results." Keeping their identities hidden, however, fueled distrust. Endesa's so-called best forest nursery is Mininco, which belongs to the same family business group that owns Colbún, the minority partner in Hidro-Aysén. And the purported best forest engineer came from INFOR, responsible for replacing native forests with industrial plantations. Mininco would generate additional profits for HidroAysén's parent companies through the sale of thousands of saplings. As the architect of forest plantations, INFOR was yet another example of state funding promoting private profits, with negative ecological impacts. These relationships therefore appeared symptomatic of a tight-knit economic system that favors a few big players to the detriment of society at large. Keeping it secret only heightened the distrust.

Conaf, the national forest agency responsible for verifying HidroAysén's EIA, also dismissed Endesa's Pascua reforestation as a "legal requirement," not an experiment. But like Endesa's engineers, the official in charge failed to grasp the significance of Piper and Fajardo's criticisms. He asserted, "No one can say right now how the reforestation plan should be done, with what methods, trees, management protocols, etc. There is no experience with this. ... No one can say what is best. ... The companies have to do 'trial and error' to produce this information. ... Without experience, there is no evaluation."[41] In other words, he did not see a role for science in producing the kind of systematic, replicable information that could provide a more secure foundation on which to design a reforestation plan for Patagonia.

Moreover, he remarked, there was nothing the agency could do about Endesa's plan to plant trees on steppes rather than forest soils because the EIA rules require only that a company plant on "similar soils." He professed to have no capacity to interpret what similar soils means. His job was only to ensure HidroAysén met all applicable laws: article 11 of law 19.300 says that loss of forests is a significant impact; article 103 of decree 95 requires that this loss be compensated with reforestation; article 102 of decree 95 and article 19(g) of DS No. 93 (2008) require that the company submit a plan for approval; and the EIA Agency's guidelines explain how to prepare this plan. On page 53, the guidelines specify that the company must reforest with the same type of forest (as defined in article 2(26) of law 20.281) on similar soils (as defined in DS No. 93). Yet when I looked for DS No. 93, I could not locate the article that requires similar soils, and none of the participants in this event could provide any further information. In addition to hiding behind the rules, this official challenged CIEP's credibility. Sipping coffee from a mug with CIEP's logo on it, he shrugged and said, "CIEP is not a government entity, so it relies on external funding." He concluded, incorrectly, that CIEP had in some way participated in the baseline studies and therefore was, in his view, untrustworthy.

This episode illustrates the constantly shifting boundaries of science in Chile, even among a small group of individuals. The experiences of Piper and Fajardo speak to the special role scientists with a deep connection to place, both in a geographic and cultural sense, can play in these kinds of disputes, and some of the conditions that can facilitate scientific engagement instead of self-censorship. The equivocal role played by the Conaf staffer stands out; the official blithely blurred the boundaries between scientists and consultants, and failed to grasp how or why science could be useful in his job. The state organizations that should have been receptive to the forest ecologists' criticisms did not take science seriously, either for what appear to be political reasons (e.g., the regional governor who ignored their memo) or arguing legal impediments (e.g., Conaf).

Baselines versus Impact Evaluations

In other areas, however, Conaf held Endesa to a high standard. For example, it insisted that Endesa improve the forest baseline studies. The original studies Endesa submitted, executed by university scientists, used aerial methods to estimate the types of forests that the dams would submerge. This methodology is not contemplated in the EIA Agency's rules regarding forest cover estimates. After significant effort, Conaf staffers forced Endesa to repeat the studies with the legally sanctioned method, which required more fieldwork (which is expensive) but was more accurate. The mismatch between Conaf officials' apparent lack of concern regarding the afforestation plan's potential effectiveness and its high concern over how the forest baseline studies were done points to an irony in EIA practices in Chile: although industry staffers consider baselines to be like a photograph in which "there is not much to discuss," state officials spend most of their time reviewing and discussing baselines, not impact evaluations or compensation plans (like the proposed afforestation plan).

Reviewing baselines is a better use of state officials' time because baseline studies have no quality roof: rules and regulations may specify the format of a report, but cannot specify what exactly to study because, by definition, baselines register a locality's nature. Baseline studies are resistant to standardization; they should represent the combination of ecological and social conditions that characterize a specific locale. As a result, baselines give bureaucrats some room to raise performance standards above that required by environmental regulations. In addition, baselines provide a foundation for legal accountability. Demonstrating changes from the recorded baseline is the best proof of environmental harms resulting from an industrial project. This was the lesson many learned from the three prior conflicts discussed in this book and a driving force behind state officials' "insatiable appetite" (as an Endesa official characterized it) for more information.

By contrast, though industry and consultants described impact evaluations as messy, strategic, and political, in practice these are highly regulated. Article 11 of the environmental framework law (19.300) defines what counts as an environmental impact that triggers the need for an EIA, decree 95 specifies how to compensate these legally defined impacts, and multiple guidebooks specify when to issue specific permits. Hence, the most political part of an EIA for consultants, companies, and scientists is for state officials the most legally circumscribed. This circumscription works to reduce state officials' discretion in EIA decisions.[42]

Consider, for instance, how the Water Agency approached the problem of the dam. Endesa proposed to operate HydroAysén's dams using a technique called hydropeaking, whereby water is stored in the reservoir when electricity consumption is low and released during peak hours to maximize profits. On the Baker River, hydropeaking would have caused the water level to fluctuate between one to three meters twice daily, continuously exposing and inundating the river's banks, causing soil erosion, trapping fish and animals, and endangering navigation. A consultant explained that hydropeaking would have "produced summer and winter conditions every day, twice a day. This produces total confusion to animal and plant species that depend on the river."[43] Over time, the river's morphology would have changed (Endesa 2008, Anexo D, Apéndice 4).

Endesa and the Water Agency each hired scientific teams to study these issues. Endesa hired a specialized consulting agency, CEA, while the Water Agency hired EULA, known for its expertise in water-related environmental issues. CEA and EULA became locked in a rivalry to produce more and better information for the baseline studies. Participants in these events describe it as a slow game of strategy. The Water Agency demanded more information to improve baseline studies, while CEA pushed back. The Water Agency's efforts forced Endesa to release a lot of new information through Hidro-Aysén's EIA. For months, the agency argued that Endesa needed to submit sediment samples, and the company replied this was impossible (the Baker's current was too strong for sampling). It seemed EULA and the Water Agency had HidroAysén in a checkmate, but Endesa bested its opponent in the last round. At a meeting to resolve the issue of sediments, the Water Agency brought one US-based expert and Endesa brought several, who successfully argued in favor of the company. This episode ended the Water Agency's efforts to improve HidroAysén's baselines and shifted the agency's attention to the permits it would issue based on the impact evaluation portion of the EIA.

In Aysén or Santiago, no one had much to say about HidroAysén's impacts. When I asked about the dams impacts on navigation, a staff person at an HidroAysén open house, set up to answer citizens' questions about impacts, answered that "here in Aysén, we make do with what we have."[44] Many Ayseninos shared this attitude. For example, two men who earned their living driving boats on different sectors of the Baker River did not know if the river would be navigable after the dams were built; they would wait and see, and adapt to what came. How impacts were evaluated in the EIA stoked this complacency.

HidroAysén's EIA ranks impacts on a scale of zero to one hundred (Endesa 2008). The top impacts, the report says, would result from the inundation of habitat. How this ranking is achieved indeed appears messy and strategic. The ranking formula privileged the protection of well-preserved yet rare elements, like a native forest grove, over those used by humans for fishing or navigation, such as rivers or the seashore. At other times, however, the formula privileged changes over space rather than those that occur over time, or vice versa, without an explanation. The report is vague about who or what determines whether an impact is positive or negative; in this case, the consultants conclude that Aysén would have benefited tremendously from new reservoir lakes. Finally, the EIA describes the impacts in different terms than the public uses. For instance, the report does not use the term hydropeaking, and discusses its impacts as isolated or fragmented phenomena. Finally, the impact evaluation was sloppy; some impacts to the Baker River, for example, were ranked on a zero to ninety scale because a variable (reversibility) was measured out of two and not three points. The result of this reporting strategy is to obscure the impacts that would most affect Ayseninos and center the public discussion on how to get baselines right.

Getting to a Decision on HidroAysén's EIA

Soon after HidroAysén's EIA was approved, I asked a former director of Conama what he thought HidroAysén's most significant impacts would be. I hoped he would offer a list of major concerns: changes to the river's morphology, water quality, risks of flooding, glacier-related risks, industrial hazards, social impacts from an influx of workers, and so on. I should have known better. In months of asking this question of scientists, consultants, Endesa staff people, and activists, no one had offered a straightforward answer. This former state official, now a university professor, answered instead that there was no information about HidroAysén he could trust.[45] The EIA process he had once helped to implement had failed to generate credible public information. Whereas Endesa felt it had gone out of its way investing in science to create a good, credible project, political leaders, professors, scientists, and activists disagreed. As the moment of decision regarding HidroAysén's EIA approached, the country was divided over the project's merits, but also over what the decision should be based on. Many worried that political support for business and industry would again subvert what should have been a technical decision, or alternatively, a political one based on a shared, collective interest.

The weeks prior to the EIA Agency's decision were tense. In a disjointed, broad-ranging debate, opinion leaders and politicians laid out competing interpretations of what was happening and what should happen. Several criticized the state and government for not having a long-term, guiding vision regarding energy, and instead leaving such important decisions to the market (Mansuy 2011; Mosciatti 2011; Navarrete 2011). In response, former environment minister Ana Lya Uriarte proposed that legislative decree rather than a technical process like the EIA should be utilized to approve projects with such strong irreversible effects like HidroAysén (Ex ministra 2011). Others alleged that illegalities had been committed during the EIA evaluation. Congressperson Gabriel Silber told the press, "There have been many administrative irregularities and possibly wrongdoings during this [EIA] process. We think all kinds of pressure have been applied to break environmental institutions to approve the project.... They have tried to do environmental fast track, tampering with reports and ignoring technical opinion" (Carmona 2011). In response, many legislators and activists advocated for making the evaluation process "more technical." This was the spirit of comments by Patricio Rodrigo, director of Patagonia without Dams, made on national television.[46] Yet others saw in all this evidence of a "democratic deficit" where institutions, due process, and rules were insufficiently respected (Soto 2011; Aguila 2011). Yet another group highlighted state agencies' legal, infrastructural, and epistemic weaknesses compared to the position of powerful companies (Mena and Rivera 2011).

Scientists also criticized the decision to approve HidroAysén's EIA. Some directed their concern at the project versus the process. For example, a University of Chile engineer advocated for "sacrificing" the Pascua River, where few live, to "save" the Baker, which is central to many Ayseninos' lives (McPhee 2011). But most scientists directed their criticisms at the EIA process and the role of science within it. A University of Concepción engineer circulated a twenty-page critique by e-mail denouncing state officials' limited scientific training and lack of access to basic monitoring data; dishonest scientists who produce poor-quality and biased baseline studies; and a regulatory framework in which regulators cannot modify vital aspects of a project, like location, design, and operating method (Meier 2011). Other scientists left comments to an article about HidroAysén in the British journal *Nature*; there they complained that "there is no real science behind the approval of HidroAysén," "if the studies were exhaustive, then, why not expose the knowledge in a form that can be widely discussed?" and "the lack of environmental experts in the approval of the hidroAysén proyect [*sic*] generates distrust in the citizenship."[47] A group of ecologists sent a letter to the editor of a national daily newspaper criticizing the government for failing to consult with its "personnel of excellence," including scientists at CIEP. "We reject HidroAysén and stand in solidarity with the demonstrators," the group wrote, submitting the letter "together with foreign colleagues" while attending a meeting at the prestigious Cary Institute of Ecosystems Studies, Millbrook, U.S. (Armesto et al. 2011).

Implicit in the scientists' criticisms are two assumptions: that more and better science would lead to better EIA decisions; and that the state should take an active role in science and energy policy. The University of Concepción engineer, drawing on the final report of the World Commission on Dams, saw increased state regulatory power as the solution to the EIA "abuses" he outlined. The state should directly control baseline studies, and set the terms of reference between companies, consultants, and scientists. This solution resonated with scientists and state officials inclined to believe in the authority of science. For instance, one state official based in Aysén complained bitterly, "The government has no interest in knowing things that may raise questions about some projects" and agreed the solution lay in reforming the state.[48]

But many activists and elites took an entirely different position, in which personal responsibility and legal accountability were far more important than scientific evidence. Just a week before the EIA Agency's decision, some activists launched a campaign imploring state officials to "vote with their conscience" to protect Patagonia's future. The campaign targeted the eleven officials who sit on the committee that votes on EIA projects. The activists posted these officials' names, office contact information, and photos on a large board in Coyhaique's town square. A more aggressive version circulated on the Internet, with the caption "Wanted. ... For wanting to sell Patagonia." The campaign asked citizens to write to these officials, who are not elected but rather work for the state, to ask them to reject HidroAysén.

Reactions to the campaign illustrate the limits of Chile's democracy. Business leaders called it aggressive, and political figures denounced it as putting state officials under "undue pressure." Legal experts described the divulged contact information as private, although it was all work related (Moya and Cabello 2011; Celedón, Moya, and Esturillo 2011). An assistant regional governor defended the procedural legitimacy of the EIA, saying, "The legality and objectivity of the process must be respected, and members of the evaluating committee should be free from such pressures" (Una estrategia condenable 2011). Targeted officials were upset at having been made the villains, as if they were responsible for the region's future. According to one, "There is a lot of confusion about this. We have the responsibility to vote on this project's EIA because of our positions and environmental laws. We are not, as they say, responsible for the future of Patagonia" (Seremi 2011). Instead of a sense of responsibility to citizens and future generations, the EIA Agency's regional representative explained that the committee's vote could only be based on legal considerations:

> We do our work remaining attached to what the law dictates. We are governed by public law that tells us that we can only do what the law allows us, not what we are forbidden from doing, as is the case with the private sector. Everything we do occurs within this regulated framework, which is why rather than voting with our conscience we must vote following the letter of the law. (Actores relevantes 2011)

A few months earlier, the agency's previous director similarly explained to me, "We do not do anything outside the EIA's rules and regulations or the law. We want to narrow things down, so nothing is subjective in any sense. We want to narrow the gaps. Instructions, guides, [and] admission tests—we want these to be more regulated."[49] Across state agencies, some state officials illustrated these ideas with sports metaphors. "My job is to draw the lines on the soccer pitch," make sure the lines are respected, or be "the net on the tennis court."[50] In short, in terms Milton Friedman would appreciate, many state officials evaluating information in the EIA strived to be umpires.

Conclusion

HidroAysén was a project fit for a high modernist empire state that failed under umpire state conditions. Seen this way, the project's demise sheds light on the limits of a type of state that aspires to be a neutral referee in a competitive game, as Friedman (1962) described in *Capitalism and Freedom*. HidroAysén was a project of empire in the sense that it required enormous capital, material, and intellectual resources to be built, with its complexity only heightened by the rurality and remoteness of Aysén. Until recently, only states had the capacity to marshal the resources and resolve necessary to carry out such projects. But under umpire state conditions, such a project is built if it meets a corporate bottom line. For advocates of the umpire state this is positive; supporters of neoliberalism were critical of large state bureaucracies that spent public tax dollars as if a collective interest existed. For many Chileans who opposed HidroAysén, however, such a large, transformative project cannot be justified absent a shared vision for society. The social and environmental costs were too high given that the benefits would accrue to a small group of business elites.

HidroAysén thus reveals the limits of the umpire state. Elites in industry and politics became concerned that the project's failure pointed to the upper ceiling of Chile's economic activity. After the project's demise, Hernán Salazar, Endesa's former CEO, said, "Our institutions proved to be 'too small' for us" (Pizarro 2014). In other words, Chile's democratic institutions were too focused on short-term, legalistic goals and too disconnected from communities' concerns to legitimate HidroAysén. Throughout, Chile's executive government clung to the notion that EIA decisions had to be made according to the rules. After HidroAysén was approved, the minister for energy, Laurence Golborne (2011), confidently declared, "Every project that meets all applicable laws is good for Chile." Three years later, when HidroAysén's EIA permit was revoked, members of Bachelet's government similarly argued that their decision was based on the rules; HidroAysén, they contended, had not met the applicable rules because its EIA fell short of requirements regarding the quality of information about hydropeaking along with its impacts, such as on wildlife and amphibians, and obligations regarding relocations.[51] In short, like the previous conflicts discussed in this book and despite all the work that went into HidroAysén's EIA, many Chileans saw the decisions to approve and then reject the dams as political, while politicians of all parties scrambled to justify each decision as technical—that is, based on legal terms.

This commitment to rules is a commitment to signal to foreign investors that Chile is a good place to do business because its institutions can be trusted: decisions are predictable, not based on political whim. By contrast, in other prominent EIA decisions the role of politics is made evident. In 2015, for example, US president Barack Obama rejected the controversial Keystone XL pipeline that would have transported oil from Canada to the Gulf coast. A long EIA process thus ended at the president's desk, and he personally announced his decision on national television. Whether or not Keystone XL met all applicable laws, President Obama rejected the project based on his assessment of the national interest; in this case, the need to fight climate change outweighed the pipeline's purported economic benefits.[52]

The HidroAysén case illustrates instead how efforts to deny the role politics plays in such decisions can weaken the legitimacy of EIAs. The original goal of EIAs was to force state agencies to undertake a "fine-tuned balancing analysis" of environmental versus economic goals. As defined by a US circuit court in an early EIA case, a "good" EIA decision required state agencies to proactively and thoughtfully consider environmental values at every stage of the evaluation process, above and beyond legal formalities. This kind of reflexivity, however, was erased from the HidroAysén case when officials and politicians justified their decision by appealing to the rules as if these determined Chile's national interest. As officials and politicians tried to marginalize politics from their public statements, energy and development policy in Chile became highly politicized. In the two years after HidroAysén was approved, five different officials occupied the top spot at the energy ministry. In 2012, Aysén province experienced a social movement that cut the region off from the country for three weeks and led Colbún, Endesa's minority partner, to withdraw its financial backing from HidroAysén. Distrust of EIAs also grew, leading to new reform efforts that are discussed in the book's conclusion.

Reflexivity, though, was not entirely absent from HidroAysén's EIA studies and evaluation. State agencies and officials in practice played a mix of roles throughout the process. Sometimes officials worked to improve baseline studies, but at other times they selectively circumscribed or ignored scientific knowledge. At times they sought to hold HidroAysén to a high standard, and at others they aspired to neutrally enforce the rules. Officials occasionally, but not always, supported the official narrative that HidroAysén would fulfill Chile's energy needs. State agencies pursued a number of strategies to cope with the shortage of staff, time, and authority needed to improve the project in response to Ayseninos' demands. The case thus highlights that the umpire state model was not universally subscribed to. Rather, different visions of the state coexisted, including within the state itself. But in all these visions, state agencies depended on external sources of data and expertise. Each office collected little environmental information directly and employed at best just a few staff people with specialized training.

Negotiating the credibility of EIA science as technical or scientific, as opposed to political, was therefore of utmost importance for the legitimacy of the entire process. Yet state officials were infrequently involved in those efforts. Usually, scientists, consultants, and industry staff negotiated the boundaries of science in ad hoc, informal spaces, sometimes exposed to intense public pressure (as when CIEP was ejected from HidroAysén's

EIA, because its reputation for objectivity and knowledge of place was too weak to sustain its participation in this controversial project). Scientists did not fare well in these credibility contests. To differentiate it from consulting work, science was narrowed down to descriptive tasks (e.g., baseline studies). In doing so, scientists appeared to the public as both irrelevant—their work circumscribed to a small part of the EIA—and unethical, as they used baselines to distance themselves from the downstream consequences of their work. Scientists wishing to break these patterns, like the forest ecologists, had no opportunities to publicly perform their impartiality. How to give the HidroAysén decision a technical foundation was therefore complicated, not least because technical sometimes meant scientific, as when Endesa hired the nation's best scientists for the EIA, and at other times meant according to the rules, as in some state officials' efforts to bring closure to this controversial project.

Conclusion: Statecraft and Neutrality Revisited

Since the 1980s, the Chilean state has been led by umpire state ideals, which explains why the new democratic governments have not invested in scientific capacities. These capacities include labs, monitoring and research equipment, and the training and employment of experts who historically have been used, as described by James C. Scott and others, by high modernist empire states to exert their power over citizens and nature. In Chile, the decision not to invest in science was not a consequence of resource poverty but instead the logical result of a neoliberal ideology in which the subsidiary principle reigned supreme: the state can only do what it is expressly allowed to do, while the private sector can do anything not explicitly forbidden. Chile's umpire state is enacted through state agencies that are limited in scope and analytic capacities as well as science that has been turned into a market commodity and low-status activity. This state, however, cannot convincingly balance competing interests or values. In the specific case of EIAs, it cannot legitimate large industrial projects or create transparent foundations for accountability: documents might be made available, but the logic underpinning public policies and decisions remains obscured by claims of uncertainty and political posturing. The distrust of data and scientists reinforces distrust of state agencies. Neither scientists nor officials can persuasively perform their credibility to society.

The umpire state is not a prescribed list of policies leading to a precise end goal but rather an ideology that exists in tension with alternative ways of thinking about the state. In Chile, even state officials sometimes pushed back against umpire state ideals, such as by introducing new concepts into EIAs or demanding more baseline information to create stronger foundations for accountability should environmental harms occur. As an ideology, however, the umpire state neatly summarizes the neoliberal ideal of the state as responsible for organizing markets for everything—from pencils to truths. Neoliberal states not only embrace free trade policies but also regard the market as the only legitimate way of organizing human activity, including perhaps the most important activity of all: how we as a society produce knowledge, and decide together what is true and what is not. By contrast to liberal ideals around reason and science, "neoliberals have great faith in the marketplace of ideas; and for them, the truth is validated by what sells" (Mirowski 2009, 424). The umpire state ideal connects this lionized marketplace with an ordering of the state where the state guarantees individual freedom not through civil liberties or public welfare but instead by policing the rules for markets to operate, as allowed by the subsidiary principle. Compared to high modernist empire states, neoliberal reforms have delivered streamlined state agencies. But these agencies lack important capabilities, as officials are unable to produce knowledge or certify its credibility. Instead, as in Chile, what counts as science and when to use it in policy debates are wholly unclear questions, best illustrated by the trouble involved in distinguishing scientists from consultants.

In 2015, five years after Conama was replaced by an EIA Agency, EIAs were again slated for reform. President Bachelet appointed an advisory committee to improve EIAs' "social legitimacy," which she said was lacking. The committee recommended a "faster review process, with higher quality standards, and within a legal framework that gives companies, communities, and state agencies legal certainty" (Environment Ministry 2016, 7). The committee's work underscores the continued relevance of the central controversy analyzed in this book: How have Chile's democratic governments made decisions about natural resources? The four conflicts explored in these pages suggest that President Bachelet's reform committee misdiagnosed the central problem. EIAs lack social legitimacy not because the actors involved require more legal certainty (required for markets) but rather because many citizens demand that EIAs be more responsive to collectively defined goals as opposed to those of industry. Yet it is unclear who can speak for collectively defined goals in a polity where political authority is limited and scientific knowledge suspect.

Political theorist Yaron Ezrahi (1990) hypothesized that a society that ceased to value scientific and technical expertise as cultural resources for the legitimation of political authority also would cease to believe in the possibility of articulating collectively defined goals. Unbound by expertise and skeptical that scientists can record an objective reality, politicians would blur "myth" and "reality" in a rhetoric of self-promotion as opposed to championing public policies designed to achieve collectively defined ends. Such a society shares many traits with the ideal of the neutral broker state, and at the start of the twenty-first century, has gained renewed attention under the label of "post-truth" politics. Chile's umpire state predates as well as differs in style and tone from post-truth politics. Nevertheless, the umpire state ideal, post-truth politics, and Ezrahi's observations reflect a neoliberal ideology that the Chilean case sheds light on—a matter I return to in the second half of this conclusion. The Chilean case invites reflection on an ideological shift in the symbolic value of science and "truth" in politics observed in countries outside Chile.

The first half of this conclusion summarizes the main trends identified across the four conflicts: salmon farming, paper and pulp production, gold mining, and hydroelectricity. These are different industries with different ecological and social impacts, all regulated through EIAs, a process used globally to balance the economic and environmental values at play in new industrial projects. How, exactly, this balancing happens varies by country, although state agencies always play a central role. EIAs require that state agencies consider different kinds of knowledge—regulatory, scientific, and technical along with that provided by affected communities—in balancing competing values in a way that is perceived as fair and neutral. If states openly based their decision-making process on private interests, there would be no need to balance competing interests. President Bachelet's 2015 EIA reform committee, though, appeared captured by industry interests, as two-thirds of its members came from industry and consulting firms. In protest, a parallel citizens' committee composed of activists as well as elected and state officials formed to denounce the lack of representation of broader values and needs on the official committee (Propuesta Comisión 2016). The mere existence of this countercommittee confirms one of this book's main findings: that industry and consultants elicit a deep distrust among Chileans, particularly when put in charge of environmental institutions. The citizens' committee had less to say about another force undermining the legitimacy of EIAs: the role of science and expertise, without which estimates of the future ecological impacts of a given industrial project are no better than guesswork.

Routines of Governance in Chile 1: Unsettled Science

Since the eighteenth century, science has developed as a set of practices to observe and represent an objective physical world.[1] In contemporary Chile, scientists working for Endesa on HidroAysén's EIA invoked this ideal to defend the quality of their science; their baseline studies, they claimed, were simple "photographs" of an objective nature. Their efforts were the exception, however. Across the four conflicts analyzed here, scientists, industry and state officials, and activists were more likely to voice suspicion of the science contained in EIAs, driven by the conviction that industry-paid scientists and consultants are neither independent nor objective. These suspicions surrounded data capturing underwater conditions at salmon farms along with representations of streams and glaciers at Pascua Lama. Similarly, during the Valdivia crisis, every shred of scientific proof was disputed. And despite Endesa's efforts with HidroAysén, such suspicions ultimately diminished the credibility of its multimillion-dollar EIA effort.

Sometimes activists took bold actions to protest what they saw as corruption of the scientific process. In 2007, after Pascua Lama's second EIA was approved, anti–Pascua Lama activists occupied the plenary session of the joint meeting of the Ecological Societies of Chile and Argentina, raising banners that read "$ociety for Ecology" and "Ecologists Are Accomplices Plundering Nature." Over a megaphone, an activist questioned scientists' ethics: Why study nature if you accept industry funding to produce studies that facilitate industrial abuses on nature? (OLCA 2007). This was an exceptional event, but the controversies examined in this book highlight that underlying concerns about scientists as unethical, opportunistic, and untrustworthy also were shared among officials, scientists, and industry.

Conflicts over scientific independence were played out in instances of boundary work as participants in EIAs attempted to distinguish scientists from consultants. Occasionally each group was clearly demarcated, as when legislators gave scientists and consultants separate though equal voices on Conama's Advisory Council. Similarly, when it came to monitoring the waters around salmon farms, each group had specialized tasks. But more often, the boundaries between scientists and consultants were blurred. In the Valdivia case, scientists who defended Celco Arauco were accused of "acting like consultants" because they did not provide new evidence when making the company's case. By contrast, in the Pascua Lama case, some consultants promoted themselves as scientists because they worked hard to improve the project over the company's objections. In the HidroAysén case, many officials, activists, and professionals refused to accept the boundary that scientists and consultants drew between themselves, suspicious of the motives of anyone on the company's payroll.

Scientists used additional strategies to perform their independence and secure credibility, like narrowing science down to tasks of seemingly unassailable objectivity, such as keeping an inventory of glaciers. The most surprising strategy was "to reach results contrary to your funder's interests"—an approach cited by scientists participating in salmon farming and the Valdivia case. But this strategy can backfire since it gives credence to the idea that funders' interests shape the results of science. How can someone evaluating these findings discern whether they represent scientists' best efforts to capture an objective reality or an attempt to demonstrate independence by undermining a funder's goals?

These varied strategies highlight that the actors involved in each case disagreed on what counts as science, who can speak for scientific proofs, and what science can and cannot do for state officials seeking objective representations of the world to support their decision-making. Such instability reflects a general distrust in science driven by the suspicion that, as noted earlier, "he who pays the piper calls the tune" and that good science is expensive science. As best illustrated by the Valdivia case, the first adage harms the prospects of industry-paid scientists, while the latter harms those of publicly funded scientists. By contrast, scientists with foreign funding (assumed to be more generous than monies from the Chilean state) appear to be both independent of Chilean industrial interests and able to produce good science.

Across these four conflicts, Chilean scientists were fundamentally challenged by the fact that neither the market nor the state supplied them with opportunities to perform their independence, or, to once again use their terms, demonstrate that "the wife of Caesar is not only chaste and pure but also appears to be so." In other words, Chile features few institutionalized spaces that give scientists the opportunity to act as a critical community with the authority to routinely probe the technical foundations of the state's decisions, and those that did exist held limited power. Scientists on Conama's Advisory Council, for example, did not have a special authority as environmental experts to unpack Conama's proposals. Scientists in the farmed salmon sector, which provided the best opportunity for a critical community of the conflicts analyzed in this book, played this role through personal relationships and agency-sponsored workshops. But their opinions were subordinate to those of a Canadian expert, and their expertise was limited by poor access to the appropriate sampling sites and equipment. The remaining three conflicts offered even fewer opportunities for critical engagement because the scientists involved with EIAs primarily interacted with consultants and industry rather than state agencies.

Routines of Governance in Chile 2: State Agencies' (In)Capacities

In a context where the status and boundaries of science are unsettled, state agencies have developed capacities that rely on external scientific advice produced by university scientists and consultants. Agencies are consumers of data purchased through a market that, in theory, precludes state officials from developing an intimate or privileged relationship with an adviser. The champions of a market for science in Chile, such as CENMA's architects, based their arguments on conceptions of the state. As proponents of the subsidiary principle, they believed states should only engage in actions expressly allowed by law, while private entities should be free to do anything not expressly prohibited by law. What impact this might have on the quality or credibility of science was not something they addressed.

In practice, this has meant that agencies like Conama, Subpesca, or Sernapesca have focused on requesting more and better studies, as opposed to building in-house scientific and analytic capacities. Over the years covered in this study, state agencies asked companies to submit more information on their EIAs. Those that regulate aquaculture tried to improve the credibility of the data submitted to them by issuing increasingly precise sampling and collection instructions. Other capacities agencies might have were inexistent or limited. They lacked the ability to conduct unannounced inspections, require the use of best-available technologies, recommend alternatives, or prioritize environmental needs. Although fisheries and environmental agencies gained some of these capacities after reforms introduced in 2008–2009 and 2009–2010, respectively, for twenty years prior to that their capacities were geared toward approving EIAs—be it full EIAs overseen by Conama or the standardized assessments supervised by Subpesca. Moreover, the reforms have not equipped the agencies with the ability to shape the public authority of science; the agencies have not gained the right to create institutionalized spaces in which state officials might also participate in negotiating the boundaries of science, thereby fostering the critical communities needed for balancing competing values and considering different knowledges. Neither have they gained the capacity to produce their own representations of the world and render these official.

Within these constraints, state officials have pursued different strategies, revealing dissenters from Chile's umpire state. At one extreme, state officials working on HidroAysén's EIA echoed a strategy used by activists at Mehuín (Valdivia) and Huasco (Pascua Lama): they saw baselines as a weak link that they could use to disrupt EIA routines, so as to highlight a project's social and environmental injustices. In demanding that Endesa supply more information to improve baselines, officials were exercising their discretion and incorporating a local gaze into what others hoped would one day be a standardized report that treats every industrial project as well as locale the same way, and applies the same compensation to every risk or impact, irrespective of a project's size or local characteristics.

In other cases, agency staff introduced new concepts or tools to challenge claims made by more powerful actors. Thus, after it gained the legal authority to shut down highly polluting salmon farms, Subpesca regulated the market for consultants to gain confidence in the data that consultants provided and thus justify the agency's future actions. Conama staff in the Valdivia and Pascua Lama cases introduced new tools (e.g., tertiary treatment) and concepts (e.g., the precautionary principle) to improve each project in crucial ways. But these efforts were never as powerful as they seemed to be. Sometimes they backfired, as in the tertiary wastewater treatment installed at the Valdivia mill. After crisis struck, Austral scientists identified the treatment as the likely source of toxic pollutants. At other times they failed to set a precedent; for instance, at Pascua Lama, the technical foundations of Conama's EIA decision were too illogical to set a precautionary precedent, as confirmed years later by a court verdict that held Conama, not Barrick, responsible for potential harm to glaciers. Alternatively, as with Subpesca, the agency's efforts were developed for an internal audience and therefore not accompanied by measures to improve the agency's accountability with a skeptical industry and public. Occasionally such measures found support from high-profile individuals; as the events narrated in this book were winding down, Senator Guido Girardi and Juan Asenjo (2011), then president of Chile's Academy of Science, penned an editorial together clamoring for better ways to integrate expert advice into governance.

Other state officials aspired instead to set "the lines on the soccer pitch." These officials were less concerned with the scientific content of EIAs than with verifying that formalities, requirements, and standards were met. In their self-descriptions as referees, arbiters, and even inanimate objects like nets and lines, these officials embodied the state as a neutral umpire. Rather than seek to represent a region like Aysén, the public good, a vulnerable community, native forests at risk, or water quality, these state officials depicted the state as if it were itself a depoliticized, neutral agent. In this ideal, state-sponsored actions are not instrumental in the sense of providing a means to achieve an anticipated end such as environmental protection or economic growth, or a desirable mix of the two. There is no *balancing* of competing interests. Rather, in this ideal, state actions seek to facilitate a competitive "game" designed to enable players to pursue their own private agendas. The game itself, in which one player will vanquish the other, becomes the common good. The two irreconcilable worldviews in the HidroAysén case illustrate this game particularly well, pitting those who supported the dams if the projects made economic sense against those convinced they could only make sense as part of a more powerful and coherent vision of the nation's development goals. Officials, meanwhile, refereed the game, seeing both themselves and the rules as neutral arbiters.

The HidroAysén case shows that some projects are too transformative for an umpire state to legitimate, thereby placing an upper limit on the kinds of activities an umpire government can sanction while maintaining political and social stability—a conundrum that an empire state would not face because of its material, intellectual, and political resources. Nevertheless, in either an empire or umpire state, to impose projects through decision-making processes that are seen as either exclusionary or rigged is to engender conflict and distrust of elites.

Science and Environment in Post-Truth Democracies

Earth Day 2017 in the United States will be remembered not only for the usual banners celebrating a healthy planet but also for new signs reading "Make America Think Again," "Science Not Silence," and "Facts Are Our Friends." On Earth Day that year, scientists marched on Washington, DC, in defense of the scientific enterprise, under attack from the new administration of President Donald J. Trump. On taking office, the head of the EPA eliminated information related to climate change, water quality, and renewable energy from the department's website. The administration imposed "gag rules" on EPA employees and scientific advisers working for government, and proposed a budget that slashed funding for science, particularly for climate change and environmental research.[2]

The March for Science grew into "the largest global science event in history," with over a million participants in 610 marches happening from Uganda to Vietnam.[3] Chile alone saw four events. National differences aside, the values of the March for Science resonated widely. Marchers defended the notions that science should be for the common good, it is neither partisan nor at the service of special interests, it provides a solid foundation for policy decisions, and that a robust, diverse scientific enterprise can deliver human progress.

The March for Science resonated with renewed anxiety that the authority of truth in public affairs is declining—a topic that STS scholars have been writing about for some time (Ezrahi 1990; Mirowski and Sent 2008; Brown 2009; Collins and Evans 2002; Owens 2015). In the United States and United Kingdom, this anxiety has been labeled post-truth politics. The term came into widespread use after the 2016 Brexit campaign for the United Kingdom to leave the European Union and Trump's successful presidential run in the United States. These campaigns relied on tactics many now identify as symptomatic of post-truth societies: truth has become irrelevant, politicians repeat lies that have been disproven by the media or experts, and appeals to emotion and belief are more persuasive than arguments based on objective facts.[4] Notwithstanding the long histories of these tactics in politics, observers of the post-truth phenomenon argue that their use has recently intensified.

The organizers of the March for Science attempted to counter post-truth politics and declining faith in the ideal of science by defending the special status of science as a source of objective knowledge that needs generous public funding. This response, however, risks glossing over the complexities involved in connecting science and government, and ignores a wealth of social and historical research that has criticized the ways that scientific expertise has naturalized powerful interests and marginalized lay concerns. The experience of Chile recounted in this book offers foundations for a more nuanced response—one that highlights that the stakes are not just about science but also about competing visions of society that are playing out over science.

For much of the twentieth century, experts worked in alliance with the state, helping to expand its role in the economy and public life through the promise of rational instrumental actions. The Trump administration's attack on science is thus best understood as an attack on the forms of government and statecraft scientific experts have historically enabled: a "large" state capable of regulating the economy, education, health, land, and natural resources (Miller 2017). By contrast, champions of neoliberal values seek to build a state designed to sustain a market society. One of the most crucial differences between these visions of the state lies not in ideas about economic growth or trade but rather in assumptions about knowledge production. For neoliberals, and as laid out by Friedrich Hayek (1945) in a foundational text, "'the market' is posited to be an information processor more powerful than any human brain" or organization, "but essentially patterned on brain/computation metaphors" (Mirowski 2009, 435). Hence, while the welfare state relied on the wisdom of experts to devise policies to advance the collective good, neoliberal ideology denigrates expertise in favor of the "wisdom of the crowds." Chile's ideal umpire state, in which the denial of expertise and objective knowledge results from as well as reinforces the denial of the possibility of collective goals, is one manifestation of this neoliberal ideology. In Chile, the approach has resulted in state agencies that are overwhelmed with information and at pains to demonstrate their own neutrality by rigidly sticking to the rules. Among other transformations (Ezrahi 1990), these trends have consequences for the state's treatment of information and civil servants.

The possibility of alternative facts has captured the post-truth imagination, raising the potential that parallel, irreconcilable realities can coexist. Ezrahi argued that as skepticism in the ability of science to represent an objective reality grew and faith in the possibility of collective actions declined, society would no longer value coherent political programs and actions. Rather, "coherence tends to stand for pretense, untenable claims of knowledge and authority, and the unacceptable exercise of power. Incoherence, by contrast, seems to indicate humility" (ibid., 284). In contrast to the hubris of government by experts, such humility is one of purpose and knowledge as politicians abandon the belief that collective actions can result from the aggregation of individual preferences or calculations made by a central authority with privileged access to information. To believe in alternative facts is suddenly and shockingly an act of humility.

Governance that starts from the premise of alternative facts has no use for democratic procedures like deliberation, dialogue, negotiation, or consensus that seek to produce shared narratives that become legitimate by virtue of the process that brought them into being. The state has no strategic role to play in the production of information (even through democratic means like those just listed), and citizens can chose their alternative facts from the Internet. Wikipedia and Google are precisely the kind of market-based information processor Hayek envisioned (Mirowski 2009). In such a world—recalling the debates in chapter 1 of this book—myth becomes reality, and so the official Chilean Way that elites champion, describing a capable, effective state, is as real as governance done a la chilena, its popular doppelgänger. Straightforward questions, such as those left unanswered by HidroAysén's seemingly exhaustive EIA—for example, Would the Baker River still be navigable if the dams were built?—become obfuscated by posturing for and against the dams. As democratic procedures for respectful and reasoned decision-making come to seem unwieldy and pretentious, a debate prone to exaggeration and fearmongering can follow, further weakening democratic social norms.

Alternative facts present daily challenges for state agencies. When designing, assessing, or implementing a given policy, what are state officials' responsibilities when they find deceitful or misleading information? Are they responsible for fact-checking and correcting the ministers they work for, or is their duty to implement the elected government's policy program, however shaky its factual foundations? What opportunities and resources do they have to obtain the best-available information on a given issue? Information in hand, can they "speak truth to power" without fear of retaliation? Chilean officials evaluating EIAs faced these challenges daily. When they tried to push back against elected officials—who seemed unanimously committed to approving these industrial projects—they often found they had few tools and opportunities to be effective fact-checkers, or assemble the best evidence to support their misgivings. Many felt censored or afraid to speak truth to power. The response to these challenges offered by the literature on EIAs is to strengthen the autonomy of civil servants from electoral politics through clearer and stronger rules, including those related to officials' employment conditions. These measures are beneficial in any country in order to protect officials from political pressures. By themselves, however, such measures are unlikely to help officials navigate the challenges arising from post-truth or umpire state politics.

Instead, the Chilean experience suggests that state officials also require personal autonomy to exercise their discretion. An intellectually autonomous yet responsible official must be expert, honest, and free to be outspoken. Such officials must have regular access to institutionalized spaces through which they can interact with external experts, cultivate relationships based on mutual trust and respect, and hear from a plurality of informed advisers to which they are also accountable. Officials must have access to data they trust. In short, a capable yet accountable agency requires that officials routinely participate in negotiating the public credibility of scientific data and experts through institutionalized channels. By contrast, reforms geared to increase "legal certainty" in EIAs, such as increasingly standardized baselines and shorter review periods, would only disempower officials under the guise of making EIAs more technical in a legal sense.[5] Rather than increase trust, reforms that aim exclusively for increasingly precise procedural regulations would advance umpire state ideals and smother officials under an endless stream of requirements that emphasize fealty to process over results.

With their globally preeminent universities and agencies whose standards are frequently used by developing countries, the United States and European countries have more to lose by debunking expertise than a peripheral country like Chile. In the United States and United Kingdom, many commentators have responded to post-truth politics by calling for more science education, bridging the cultural divide between scientists and nonscientists, or engaging universities in society. These are undoubtedly positive steps. Yet there is a potential danger in inadvertently reifying the boundaries between science and nonscience, leading—as in Chile—to narrowing the scope of a publicly influential science.[6]

STS research has found that the credibility of science for public actions often relies on blurring the boundaries of science through carefully managed negotiations involving state officials, scientists, and different stakeholders, conducted in institutionalized spaces that allow for routines of governance to develop (Jasanoff 1990). A Chilean example of this is seen in chapter 4 with Celco Arauco's research consortia that bring together scientists with community and industry representatives to transform public concerns into an institutionalized research agenda. The consortia underscore the power of science to create social stability around conflictive projects, but raise questions about the power of industry to replace the state in the production and validation of science for regulation. What evidence would state agencies marginalized from knowledge production use to hold a company and its scientists accountable for environmental harms? Within company-controlled consortia, would the distribution of power be transparent, including the power to define what are certain and uncertain knowledge claims? Under what conditions, if any, would such consortia increase the public credibility of science done for hire?[7] It seems unlikely that better funding, such as Celco Arauco can provide, would improve the credibility of Chilean science without further changes to the market for science and the umpire state. A public discourse that acknowledges the importance of science in deliberative and plural democracies will be required to counter a neoliberal philosophy oriented toward producing a market society in which the state has reduced responsibilities for economic and social outcomes.

Appendix: Methods

Interviews: Selection and Process

I draw on material collected in interviews with strategic informants, organized into five groups (table A.1). My goal was to interview at least five people from each group for each conflict, including all individuals who played strategically important roles.

I missed some of my interview targets for a number of reasons. In some cases, the individuals involved had moved on and could no longer be reached. More commonly, I missed my target because just a few people had actually been involved in the events. For instance, even for large projects like Pascua Lama and Celco Arauco's Valdivia mill, within state agencies and companies only a handful of people might have been directly involved. That said, the conflicts varied in the number and kinds of individuals involved. More people were involved in salmon farming (where I examined the industry as a whole) and HidroAysén (which was happening in real time) than with the Valdivia mill and Pascua Lama, where I was asking about events that had occurred at least five years before my fieldwork began. Few individuals or organizations that I approached for an interview turned me down, and in all but three cases I was able to speak to at least three individuals in each "box."

To my surprise, I found that the groups were quite distinct. I assumed scientists and consultants would overlap more than was the case. Instead, more overlap existed between consultants and state officials, as it is common for Conama employees to be picked up by environmental consulting firms and experienced consultants to take higher-level positions in Conama (now the Environment Ministry and EIA Agency).

I identified individuals through several means: EIA documents, print media, and referrals (e.g., modified snowball technique). EIA documents for every industrial project built after 1997 are accessible online. These list who the Conama or EIA Agency official in charge was, and though it is sometimes difficult to find, the documents generally also list the consulting companies and scientists involved. In the case of salmon farming, I found individuals through advertisements and editorials published in industry magazines like *Mundo Acuicola* and *Aqua*.

In addition to the interviews tallied in table A.1, I also interviewed about twenty-five other informants. These included former and current directors as well as employees

Table A.1
Interviews, by Group and Conflict

	Scientists	Consultants	State officials	Industry	Activists
Salmon	5	6	4	5	1
Celco Arauco's mill	7	1	3	3	5
Pascua Lama	5	2	4	3	4
HidroAysén	11	4	9	5	3

Note: There is some double counting in this table, as some individuals participated in more than one conflict and more than one group.

of Conama and the EIA Service, environmental consultants and lawyers, and people involved with CENMA. I occasionally interviewed several individuals as a group, but I count these as one interview.

I used a semistructured interview protocol. Interviews typically ran about two hours, though some went on for much longer. In a few cases, interviewees hosted me in their homes or showed me around some of the sites in question. My questions were organized as follows:

• Details of each conflict and the informants' role in the events
• What role science played in the conflict, and whether it met their expectations
• How the quality and credibility of scientific work is communicated to others, and how conflicts of interest or concerns with quality are dealt with
• What terms like technical, science and political mean
• Their assessment of the EIA process: what worked, what did not work, and why
• The nature of their work (e.g., "What is the hardest part of your job?")
• Opinions about scientific advice in Chilean environmental politics in general (e.g., "Where do you go for environmental information you can trust?")

Following the terms of my human subjects approval, I gave every individual an informed consent form, which binds me to protect privacy. Therefore, in the narrative I do not reveal individual identities, and in particular, I endeavored to protect opinions that interviewees might find compromising. That said, I opted to use people's real names when four conditions were met: the interviewee expressed their willingness to be named (this included most of my interviewees), my interview material was corroborated by what other interviewees told me, the information I wrote about is available in the public record (newspapers, academic journals, legal transcripts, and so on), revealing the individuals' identities enriched the analysis in meaningful ways. On some occasions, however, I used pseudonyms to protect that informant's privacy. These cases are indicated in footnotes, where appropriate.

Documentary Sources

The analysis also draws on documentary sources. These vary from case to case, depending on when the events happened and what the core issues were. Below I list the full range of documents I consulted:

Public Documents
- EIA documents specific to each conflict
- EIA policy documents, including methodological guidelines and manuals developed to explain the EIA process to users
- Archival EIA documents, such as manuals from the 1990s developed by Conama to explain the process and documents relevant to the history of CENMA
- Policy documents from the Water Agency, Fisheries Agency, and Fisheries Subsecretariat explaining policies regarding stream flow, water quality, and aquaculture
- Environmental quality and emission standards
- Legislative histories, compiled by Chile's Congressional Library to keep a record of legislative debates for each law passed

Legal Documents
- The full transcript of the case *CDE (Estado Fisco de Chile) v. Arauco (Celulosa Arauco y Constitución)*
- The verdicts in several other cases filed against Celco Arauco
- The verdicts in cases filed against Barrick Gold
- A legal brief prepared by lawyers fighting Endesa, provided to me by one of the lawyers involved in the case

Scientific Documents
- Peer-reviewed journal articles by the scientists involved in each conflict and CENMA scientists
- Articles by the scientists involved in the different conflicts published in Chilean media, such as newspapers, locally edited books, and magazines, including the magazine *Ambiente y Desarrollo*, published by the NGO CIPMA
- Master's theses completed by students at Chilean universities that were relevant to each of the conflicts analyzed here
- Textbooks, where I consulted things like "carrying capacity," "acid rock drainage," "ecological streamflow," and so on

To access these documents, I used the Internet and visited several libraries as well as agency collections. The libraries at the water, geology, and fisheries agencies were most useful along with the (small) archive held by the new Environment Ministry. I also collected information about the global spread of EIAs, including documentation about the World Bank's support for EIAs, at the EPA library in Washington, DC.

Finally, throughout the book I draw heavily on Chile's news media, particularly the country's two most important newspapers, *El Mercurio* and *La Tercera*. But I draw also on two alternative sources, *El Mostrador* and the satirical paper, the *Clinic*. These have an editorial line that aligns with environmental values and is highly critical of government policy. In the case of HidroAysén, I also regularly followed two local newspapers, *El Divisadero* and *Diario de Aysén*. All quotations from Spanish-language materials in the text are my own translations.

In-Person Observation

While in Chile, I had the opportunity to attend several meetings organized by the new EIA Agency to explain to state officials, consultants, and the general public the consequences of the new environmental legislation. I was also invited to two workshops organized by NGOs that year as well as a meeting between Endesa and the families that would need to relocate if HidroAysén were built. The opportunity to be present at these meetings enriched the analysis on these pages.

Notes

Preface

1. Speech delivered by President Obama on November 17, 2016. Reported in *Time* magazine, accessed October 20, 2017, http://time.com/4575981/barack-obama-fake -news-democracy-facebook/.

2. Andy Stirling, "Science, Brexit, and 'Post-Truth' Politics," STEPS Centre (blog), July 15, accessed October 20, 2017, https://steps-centre.org/blog/science-brexit-and -post-truth-politics/.

Introduction

1. "Maria" is a pseudonym. This narrative is based on notes from my first visit to CENMA in July 2011. I have since visited CENMA four more times, and have spoken with many of its scientists and staff members. No one had a photo of the original Japanese garden, but everyone had similar narratives. I also visited three archives to look for a photo, unsuccessfully: the Ministry of the Environment, University of Chile, and JICA's Santiago offices (the latter two denied me access to their files).

2. The scholarship on neoliberalism in Chile is vast, and I selectively cite some of it throughout the book. I define neoliberalism following *The Road from Mont Pelerin* (Plehwe 2009; Mirowski 2009), an edited volume that studies the members of the Mont Pelerin Society, who self-identified as neoliberals into the late 1950s. Dieter Plehwe and Philip Mirowski demonstrate that neoliberalism is not a monolithic force. Rather, like all intellectual and ideological projects, it contains factions, contradictions, and pragmatic solutions. In Chile, neoliberalism has been present since the 1950s and proven adaptable to different political regimes (Fischer 2009; Tironi and Barandiarán 2014).

3. This includes Santiago, Temuco, Concepción, Osorno, Puerto Montt, and Valdivia and the industrial port cities of Coronel, Ventanas, Huasco, Tocopilla, and Mejillones. Based on air pollution data from Chile's Ministry of the Environment.

4. Policy studies sometimes conclude that culture matters, but leave it in vague terms such as "context" (e.g., Kolhoff, Driessen, and Runhaar 2013; Marara et al. 2011; Stein et al. 2005).

5. For a discussion on how entities like institutions or cities "think" from a coproductionist perspective, see Muñoz-Erickson, Miller, and Miller 2017.

6. Appeals to the authority of rules over the discretion of individuals (expert or not) have a long history, such as in US state agencies (Porter 1995), the ideal modern bureaucracy (Weber 1946), and Chilean political culture (Hilbink 2007). Yet this penchant for rules combined with the expansion of markets into knowledge production is a recent phenomenon. In 1990, Yaron Ezrahi described it as the rise of the "facilitator state." Similar in their goals, the umpire differs from the facilitator state in its constitutional foundations (e.g., subsidiary principle), which privilege the private sector over state actions in ways even Ezrahi did not anticipate.

Chapter 1

1. Constitutional reforms in 2005 and 2014 each removed certain legacies from the dictatorship era.

2. The Internet is full of instances of jokes associated with doing things a la chilena. For example, one group, called Buenos Para Nada Productions, made the following film about the phrase: accessed July 6, 2014, https://www.youtube.com/watch?v=wrUodRreY5w.

3. In response to President Piñera's triumphant use of the phrase, the satirical paper the *Clinic* published a cover story brilliantly titled "The Chilean Weá: de papelito a papelón" (October 28, 2010, no. 367). Weá here is written to evoke the English word "way" and play on the Chilean slang word "*huevada*," which means something silly or stupid. The subtitle refers to the little piece of paper that the miners sent from their subterranean trap and that President Piñera showed off to the world. As the president celebrated the *papelito*, it turned into a *papelón*, which is slang for "a mess," as Chileans questioned the cost of the operation and lack of regulatory oversight that facilitated the accident.

After leaving office, President Piñera took this theme of the Chilean way further in an address delivered at the London School of Economics in April 2014, titled "The Chilean Way to Development," accessed May 2, 2015, https://www.youtube.com/watch?v=vUhJUZ438j0.

4. State official 18, interview by author, April 2011. This official, then employed at the new enforcement agency, provided the author with a PowerPoint presentation with these numbers. He added that if the agency were not to enforce EIA permits, it would have little work to do.

5. NEPA's triumph over the alternative—decision-making rooted in internal modes of expertise—mirrors the rise of cost-benefit analysis by the US Army Corps of Engineers described by Ted Porter (1995). The drivers in each case are different, however. NEPA triumphed through political brokering (Milazzo 2006) rather than from the rivalry between weak agencies (Porter 1995). Whether or not NEPA's success owes something to the rise of the mechanical or rule-oriented modes of objectivity that Porter identifies is a question for historians of the United States.

6. By contrast, this is not the case in Latin America (Carruthers 2008).

7. Judge Wright delivered this statement in his verdict on Calvert Cliffs' Coordinating Committee, Inc. v. United States Atomic Energy Commission (1971).

8. I am not suggesting that procedures are not important. They most certainly are. It is also necessary to distinguish administrative procedures from environmental quality laws (e.g., caps on emissions as well as air and water quality standards) and land use planning laws (e.g., that restrict certain kinds of activities and construction in specific spaces). When Chilean officials talk about "the rules," they are referring to "administrative procedures," not quality standards, emission laws, or land use planning policies, which, if stronger, would likely lead to significant improvements in environmental outcomes.

9. For a case in East Africa, see Thompson 2004. For one in Chile, see Schaeffer and Smits 2015.

10. See also Centellas 2010, 2014; Shrum and Shenhav 1995.

11. For a broader discussion of the limits of global policy interventions, see the introduction in Mitchell 2011.

12. Studies of critical communities and civic epistemologies are interested in identifying patterns and routines of governance. This differs from related studies in environmental politics that focus on understanding policy change resulting from social movements, enabling or blocking networks, or policy diffusion (e.g., Duit 2014; Hochstetler 2011; Kinchy 2012; Steinberg 2001). Notably, this latter group of studies tends to emphasize that Latin American institutions are especially mutable compared to their counterparts in the United States or Europe (Hochstetler and Keck 2007; Steinberg and VanDeveer 2012). The experience of Chile, marked by massive political disruptions from democracy to dictatorship, certainly speaks to this mutability. Nevertheless, as this book shows, mutability is not the whole story.

13. Of relevance here, a leader of this protest attributed the problem to a cultural crisis (Fajardo 2015).

14. Compared to other systems, some notable features of Chile's EIAs are, for one, that the law defines "significant impacts." Unlike in other countries, state officials have little discretion over which projects need to submit an EIA. Second, EIA decisions

are made by regional ministerial representatives—not an independent council (as in the United States) or environmental agency (as in Ireland). Third, Conama or the EIA Agency cannot introduce significant changes to a project, like moving its location or requiring the use of best-available technologies (as in the United States). On Chile's EIAs, see Espinoza 2007. On the 2010 reform, see Barandiarán 2016.

15. Between 1997, when EIAs became required, and 2010, the SEIA rejected 47 out of 854 EIAs. But it only approved 594 (70 percent) of submitted EIAs. The rest were withdrawn or deemed inadmissible. These numbers are in line with experiences in the United Kingdom and the Netherlands (Glasson, Therivel, and Chadwick 2012; Pope et al. 2013).

16. Across the globe, consultants working in the for-profit private sector are heavily involved in preparing EIAs. Some countries have implemented quality control measures, such as keeping a list of certified consultants. But many other countries, including Chile, have not implemented any regulations to certify consultants (Glasson, Therivel, and Chadwick 2012).

Chapter 2

1. For a detailed discussion of the origins, evolution, and uses of the US social contract for science, see Guston 2000, chapter 2.

2. Nonetheless, until the 1960s the museum was the primary place of training and practice for Chile's incipient community of ecologists (Jaksic, Camus, and Castro 2012).

3. Croxatto's work eventually contributed to new treatments for diabetes and hypertension that have helped millions in Chile and globally.

4. Activist 9, interview by author, October 2010.

5. The origins of Chile's environmentalist movement merits further research. According to Fabián Jaksic, Pablo Camus, and Sergio A. Castro (2012), Chile's first modern environmentalist was Rafael Elizalde, a state official who in 1958 published the book *La sobrevivencia de Chile*, denouncing environmental degradation. In 1970, an elderly Elizalde set himself on fire, seemingly to protest indifference to environmental degradation. Elizalde's death by fire was ruled a suicide.

6. During the 1980s, CIPMA published the journal *Ambiente y Desarrollo*, and used it to record what happened at these meetings and invite further reflection. The material in this paragraph comes from notes published in *Ambiente y Desarrollo* 3, nos. 1–2 (1986).

7. Activists 6 and 9, interviews by author, October 2010. Similar statements are collected in the notes section of *Ambiente y Desarrollo* 3, nos. 1–2 (1986); Jaksic, Camus, and Castro 2012.

8. Consultant 13, interview by author, June 2011.

9. LH stands for "legislative history," a document compiled by Chile's Congressional Library to archive congressional debates around specific laws.

10. Scientists 19 and 28, interviews by author, June and July 2011. The functions of the Advisory Council changed significantly after the 2009–2010 reforms (Barandiarán 2016).

11. This is a problem elsewhere too. See Goldman 2005.

12. State official 28, interview by author, November 2010.

13. Consultants 1, 3, 4, and 10, and state officials 19, 20, 21, 24, 25, and 28, interviews by author, November 2010–July 2011.

14. Consultant 13, scientists 19 and 29, and industry staff 17, interviews by author, June–July 2011. These statements were confirmed in a letter provided to the author by an interviewee from Jaime Lavados (chancellor of the University of Chile) to Enrique Silva (minister of foreign relations) detailing the creation of CENMA and noting JICA's opposition to putting the center within the university, dated June 23, 1993.

15. Industry staff 17, interview by author, June 2011. This information was confirmed by meeting minutes signed by Adriana Hoffman (Conama), Juan Escudero (CENMA), Luis Bahamonde (University of Chile), and Jumpei Watanabe (JICA), extending JICA support for two years (2000–2002), and specifying the commitments of Conama, CENMA, and the University of Chile, dated August 11, 2000.

16. The author obtained the budget information through a public information request. This information was further confirmed in two letters, given to the author by an interviewee; one from Luis Riveros (University of Chile) to Pablo Ulricksen (CENMA), complaining that Conama's contribution to CENMA was cut to one-third of CENMA's needs, dated November 17, 2000, and another from Riveros to Adriana Hoffman (Conama), complaining of budget cuts, dated November 2, 2000.

17. Scientists 19, 30, and 31, interviews by author, June 2011 and September 2014. In 2013, Victor Perez (University of Chile) sent a letter, dated July 9, 2013, to CENMA informing it that the university planned to close the center. CENMA's workers' union replied with a letter protesting this decision, dated August 6, 2013. Both letters were provided to the author by an interviewee. CENMA survived those efforts but in late 2017 new rumors that it would be closed surfaced. I was unable to confirm these with CENMA staff, who did not respond to email or answer the telephone.

18. State officials 21 and 25, and consultant 13, interviews by author, June–July 2011. I met only one state official (24) who was enthusiastic about CENMA's work.

19. Scientist 19, interview by author, June 2011. See also CENMA 1999; Prof. Eugenio Figueroa 2003.

20. Industry staff 17, and scientists 19, 29, and 30, interviews by author, June–July 2011 and September 2014.

21. Scientists 30 and 31, interviews by author, September 2014. See also public tenders 2009 #757–189-LP09 and 2010 #757–92-LP10, accessed October 27, 2017, www.mercadopublico.cl. An interviewee provided the author with a letter from Italo Serey (CENMA) to Maria Ignacia Benitez (minister of the environment), complaining about the air quality monitoring contract as well as denouncing the start-up's lack of experience and qualifications, dated May 2010.

22. The Ministry of Health later switched providers again, to the Meteorological Service of the military's Air Traffic Control Division.

23. Scientists 19, 30, and 31, interviews by author, June 2011 and September 2014.

24. Scientists 29 and 30, and consultant 13, interviews by author, June 2011 and September 2014.

25. I am thankful to scientist 30 for this observation.

26. Some state-run labs have been privatized (e.g., Mining and Metallurgy Research Center, which in 2016 the government re-activated) or driven into the arms of industry (e.g., INFOR). See also Bustos 2015.

27. Scientists 29 and 30, interviews June 2011 and September 2014.

28. State official 22, interview by author, June 2011. For an economist's textbook on public policy, see chapter 15 of Lee Friedman's *The Microeconomics of Public Policy Analysis*. Corfo's tender has wandered even further from its original intention because it now supports companies' information needs, not those for regulation by the state. See "Bienes Públicos para la Competitividad," Corfo, accessed October 27, 2017, www.corfo.cl. A similar shift with similarly difficult consequences for laboratories is occurring in the United States (Mirowski 2011; Berman 2012).

29. Consultant 2, interview by author, October 2010.

30. Activist 8, interview by author, June 2011.

31. Scientist 28, interview by author, July 2011.

32. Industry staff 11, state officials 15, 17, 18, 20, and 25, scientist 30, consultant 13, and activist 1, interviews by author, November 2010–July 2011 and September 2014.

33. SustenTank (2011) reached a similar conclusion.

34. This is true of the debates in the early 1990s and reforms in 2009–2010 (Susten-Tank 2011; Barandiarán 2016).

Chapter 3

1. Chapter 2 discusses the historical importance of CIPMA and its conferences. The following material is based on my fieldwork notes taken during CIPMA's eighth conference, held in Santiago on April 14, 2011.

2. Farms submit a *declaration* of environmental impacts, which undergoes a limited review compared to EIAs. Although the impact studies were shorter, the baseline and monitoring requirements made up for it by requiring the kinds of information that EIAs have, but declarations do not. This system avoided overwhelming agencies with huge EIAs for similar operations and held all farms to the same reporting requirements. As we will see in chapter 6, many in Chile believe that EIAs as a whole would improve with standardized baselines.

3. Chilean lawmakers are not alone in these struggles. The concept has a long history of mixed uses (Sayre 2008), and as used by waste scholar Max Liboroin (2013), forces a radical rethinking of economic growth.

4. Scientist 6, interview by author, April 2011.

5. Scientist 7, interview by author, January 2011.

6. Dollar figures are approximate and were converted from Chilean pesos at the 2007 exchange rate.

7. State official 6, interview by author, May 2011.

8. Several interviewees complained that scientists do not want to work on problems that are relevant to industry and therefore needed incentives to force them to do so (scientist 7, consultant 7, and state official 6, interviews by author, January and May 2011).

9. State official 8, interview by author, May 2011.

10. Scientists 7, 8, and 10, interviews by author, January and May 2011.

11. Scientist 8, interview by author, May 2011.

12. Scientist 6, interview by author, April 2011.

13. Consultant 5, interview by author, May 2011.

14. Industry staff 4 and 5, interviews by author, May 2011.

15. Scientist 9, interview by author, July 2011.

16. See IFOP's website, accessed December 12, 2015, http://www.ifop.cl/?page_id=415.

17. Scientist 10, interview by author, May 2011.

18. At the time of the ISA virus outbreak, the salmon in the first infected farm were already quite sick and this increased their susceptibility to the influenza virus. That

said, there is no surefire scientific indicator for a pathogen like ISA (Godoy et al. 2008; Soluri 2011).

19. Consultant 9, interview by author, May 2011.

20. State official 7, interview by author, June 2011.

21. Consultants 5, 7, and 9, interviews by author, May 2011.

22. Consultant 6, interview by author, May 2011.

23. State official 7, interview by author, June 2011.

24. Consultant 6, interview by author, January 2011.

25. Consultant 9, interview by author, May 2011. Interviewees from all sectors were nearly unanimous in expressing distrust of the data consultants collected. No one admitted to manipulating the data themselves, but many pointed the finger at others.

26. Consultants 7 and 10, interviews by author, January and May 2011.

27. Consultant 7, interview by author, May 2011.

28. The final legal text appears to have relaxed this standard to two infractions in a two-year period (Fuentes Olmos 2014).

29. State official 7, interview by author, June 2011.

30. Consultant 9, interview by author, May 2011.

31. Scientist 6, interview by author, April 2011.

32. The following material is based on Subpesca 2008b.

33. Between 2008 and 2010, legal changes were introduced to allow the government to move these kinds of concessions (Fuentes Olmos 2014), but in my interviews this possibility was never discussed as something that might actually happen.

34. State official 7, interview by author, June 2011.

35. State official 5, interview by author, May 2011. José is a pseudonym.

36. Scientist 6, interview by author, April 2011.

37. State official 5, interview by author, May 2011.

38. Consultant 7, interview by author, May 2011.

39. Scientists 7 and 8, interviews by author, January and May 2011.

40. Scientist 7, interview by author, January 2011.

41. Scientist 6, interview by author, April 2011.

42. State official 8, interview by author, May 2011.

Chapter 4

1. Interview with Roberto Schlatter, reproduced in the documentary *Ciudad de Papel* (Jirafa films, 2007). For a video of a swan suffering neurological damage caused by heavy metal poisoning and starvation, see https://www.youtube.com/watch?v=Cg6hX0ADd3M.

2. State official 28, and scientist 25, interviews by author, November 2010 and January 2011. President Eduardo Frei was a vocal supporter of the project.

3. Concerned Valdivia residents objected to the mill's location from the start (Escaida et al. 2014, 66; Sepúlveda 2016, 157).

4. State official 28, interview by author, November 2010. These kinds of statements are not unique to this case. For similar assessments of Conama in the Pascua Lama conflict, see chapter 5. The environmental framework law capped Conama staff at sixty-two (Barandiarán 2016).

5. Activist 11, interview by author, January 2011. The Mehuín activists averted a case of environmental injustice, piling waste on an indigenous community (Nahuelpan 2016). Far from being antiscience, Mapuche residents and their nonindigenous allies had deep scientific and experiential knowledge of the place. What they opposed was a sham EIA evaluation done at their expense (Skewes 2004; Skewes and Guerra 2004).

6. Local residents were again ahead of Conama on this one, as some questioned whether AOX would detect dioxins and furans (Skewes and Guerra 2004; Sepúlveda 2016).

7. Many also lost faith in Conama because the agency initially rejected the EIA outright, arguing that the studies were poor quality, but then retracted this assessment (state official 28, interview by author, November 2010).

8. Scientists 23 and 27, interviews by author, January and April 2011.

9. Activist 11, interview by author, January 2011. See also Nahuelpan 2016.

10. Scientist 24 and industry staff 16, interviews by author, May and June 2011.

11. Industry staff 16, and state official 27, interviews by author, May 2011 and November 2010.

12. Claudia Sepúlveda, who spent these years as a leading activist against the mill, later wrote a PhD dissertation about the conflict. In it, she analyzes in detail the Austral report's inductive reasoning style as well as its data and analytic gaps (Sepúlveda 2016, chapter 7).

13. CDE v. Arauco (2013) refers to the transcripts of testimony provided in the civil law case, Estado Fisco de Chile v. Celulosa Arauco y Constitución. The transcript is available at the court offices in Valdivia.

14. State official 28, interview by author, November 2010.

15. Scientists 23, 24, 26, and 27, interviews by author, January–May 2010. A few scientists felt peer-reviewed science is not relevant to social problems; scientists therefore had to choose which to produce: credible (peer-reviewed) science or relevant science (e.g., scientists 23 and 27, state official 27, and activist 10).

16. Industry staff 13, 14, and 15, and scientist 24, interviews by author, January–June 2011.

17. Scientist 27, interview by author, April 2011.

18. Alternatively, the sanctuary or wetland could be thought of as a boundary object that should have been (Leigh Star 2010) and the different names could reflect a trading zone that never was (Galison 1997). Because this is a case about regulation, however, I prefer the concept of bridging objects.

19. Scientists 21, 23, 26, and 27, interviews by author, January, February, and June 2011. One scientist likened the Valdivia mill conflict to climate change and another compared it tobacco control—two famous examples of industry-paid scientists obscuring scientific evidence of public harms to protect industry profits. Activists, meanwhile, felt strongly that all universities were "sellouts" (e.g., activists 11 and 12, interviews by author, January 2011).

20. Scientists 23, 25, and 26, interviews by author, January 2011. I was not able to confirm this information with Conicyt, which does not track rejected proposals. Whatever the realities of grant awards, this was a strong sentiment among the Austral scientists I interviewed. Additionally, some Austral scientists have written about this elsewhere (e.g., Escaida et al. 2014). After the crisis, Austral University funded some wetland research, as did the Environment Ministry in 2012–2013. Nevertheless, as of this writing, "no studies have analyzed the overall effects of the disaster on the species" (Sepúlveda 2016, 138).

21. Another way in which Chilean scientists were disparaging of "local" science is evidenced by a quote from Jaksic about the research center he directs, reported online: "We do not practice a 'Chilean ecology.' Instead we contribute to global knowledge through research" (A la vanguardia, n.d.). Here, Jaksic echoed a common idea that "Chilean" science is "applied," and less prestigious than "basic" or "universal" science (see also Kreimer 2006; chapter 2 on CENMA).

22. Scientists involved in other conflicts also mentioned this strategy (e.g., consultant 10 and scientist 6), and industry staffers believed this was a good way for scientists to prove their independence (e.g., industry staff 13, 14, and 15).

23. Scientist 23, interview by author, January 2011. For Jaksic's view of the scientific method, see Jaksic, Camus, and Castro 2012. Some claimed that hypothesis testing was the perfect scientific method, but in this case, had been corrupted by corporate funding (e.g., scientists 24 and 26).

24. Scientists 21, 24, 26, and 27, interviews by author, January and April 2011. It is worth emphasizing that these statements were made by some of Chile's most successful ecologists and biologists. Their beliefs about science and truth could not be more different from how their US peers, for example, describe the power of science.

25. In 2009, Celco Arauco received a heftier half-million dollar fine for unauthorized pollution. For a complete list of the fines Celco Arauco received, see Escaida et al. 2014, 229–232.

26. In total, the mill was closed for sixty-four days. As of this writing, Celco Arauco has not been able to find a new dumping ground.

27. Consultant 10 and state official 28, interviews by author, November 2010 and January 2011. Another blow to Conama's confidence came with the Supreme Court's decision in April 2005 to reject Austral's evidence against Celco Arauco. This was a controversial decision based on fabricated scientific evidence (Durán 2006).

28. US and EU environmental laws both use this definition. A high-ranking Chilean state official (25) pointed this out as an "original sin" in Chilean environmental law, because it effectively creates a benchmark defined by legal limits rather than real-world effects.

29. For the applicable standard (D.S. 90), see https://www.leychile.cl/Navegar ?idNorma=182637&idParte=0, accessed March 10, 2018.

30. State official 26, interview by author, August 2011. For the standard for the Cruces River, see https://www.leychile.cl/Navegar?idNorma=1084402&idParte=0, accessed March 10, 2018.

31. Industry staff 14, interview by author, April 2011.

32. Industry staff 14, interview by author, April 2011. Two million Chilean pesos at the time was the equivalent of approximately US$4,000.

33. Industry staff 13, 14, 15, and 16, interviews by author, January–June 2011.

34. This does not mean the company has not faced continued questions about its environmental performance. For information on other cases of environmental harm caused by Celco Arauco's operations, see Escaida et al. 2014; Sepúlveda 2016. And Mehuín activists continue to oppose the pipeline in their community. See No al Ducto, accessed November 1, 2017, www.noalducto.com.

35. Leading scientists involved in the case were convinced that Celco Arauco would appeal the guilty verdict. Scientists 24, 27, and 28, e-mail communications with author, August 2013.

36. Scientist 28, interview by author, July 2011. See also Zaror 2005; Sepúlveda 2016, 196.

Chapter 5

1. The name Pascua Lama was invented by Barrick when it acquired the area in the mid-1990s. Opposition to the mine has been strong among the Diaguita Huascoaltinos indigenous community, which has maintained its opposition past 2006 (Yañez and Rae 2015; Li 2017). Parallel opposition has been strong in Argentina (Taillant 2015).

2. Activist 3, interview by author, June 2011. For more on the alliances and mobilizations leading to this "unlikely victory," see Li 2017.

3. Activist 4, interview by author, July 2011. An Argentine scientist who testified against Barrick's plan similarly called the whole scheme "offensive" (Ross 2005). Jorge Taillant (2015) provides further details about Barrick's plan, including reproductions of corporate brochures advertising the operation.

4. Activist 1, interview by author, June 2011. This narrative was confirmed by activists 3 and 4, interviews by author, June–July 2011.

5. Industry staff 2, interview by author, July 2011; Junta 2006.

6. Activist 2 and industry staff 2, interviews by author, June and July 2011. The language of death is in contrast to messages about life and water (Campisi 2008; Urkidi and Walter 2011). Interestingly, the language of death was also used by activist youths against Celco Arauco's Valdivia mill (Sepúlveda 2016).

7. Industry staff 2, interview by author, July 2011.

8. Consultant 4, interview by author, May 2011.

9. Industry staff 1, interview by author, July 2011.

10. Scientist 3, interview by author, July 2011.

11. State official 3, interview by author, August 2011.

12. Activist 4, interview by author, July 2011.

13. Scientist 1, interview by author, July 2011.

14. State official 4, activists 2 and 4, and scientists 1 and 2, interviews by author, June–July 2011.

15. Pedro Bazán is a pseudonym.

16. State official 4, interview by author, July 2011. Even *El Mercurio*, a conservative newspaper that typically sides with business and political elites, criticized Pascua Lama's impacts on glaciers. See articles published on March 22 and June 3, 2005.

17. Scientist 3, interview by author, July 2011.

18. State official 4, interview by author, July 2011.

19. Scientists 1, 2, and 5, and activist 2, interviews by author, June–July 2011.

20. Industry staff 2, interview by author, July 2011.

21. Industry staffers 1 and 2, interview by author, July 2011.

22. The Junta de Vigilancia was created in 2004 to replace the historic Irrigation Association, established in early 1900s. Christina Campisi (2008) argues that this and other changes to irrigation management were implemented due to Barrick's arrival. In other words, in this interpretation, the Irrigation Association was created in this way to help ease industrial mining in the valley.

23. Scientist 3, interview by author, July 2011.

24. The consultants say they base these claims on three aerial photographs from 1981, 1997, and 2000, each with different scales (1:30.000, 1:10.000, and 1:8.000, respectively) (Barrick 2004, chapter 5, 24). However, neither the years nor the scales correspond with the photographs submitted in annex D of the same document.

25. Activist 1, interview by author, June 2011. See also Barrick 2005a.

26. Chilean activists saw the Kumtor mine in Kyrgyzstan as a cautionary tale. Between 1997 and 2007, nearly six million cubic meters of glacier ice were extracted from the site, but this was just the beginning; using traditional explosives, mine operators blasted away eleven million cubic meters of glacier ice in 2010 and another 22 million in 2011. In addition, because glaciers were discarded with other sterile rock near the mine site (to keep costs low), there are risks associated with unstable pit walls and water seepage. Furthermore, mine operators built the mine's tailings dam (toxic waste) below a glacial lake that like all such lakes, may burst its glaciated dam. Additional risks involve increased melting due to dust emitted by the mine and acid rock drainage due to having put glacier ice in the waste rock pile (Kronenberg 2013).

27. The Spanish term used in Barrick's addenda is *reptación*. I was unable to find a technical explanation for glacial "reptation." For an introduction to glaciology for nonspecialists, see Taillant 2015, chapter 2.

28. Activist 2, interview by author, June 2011. People I met while visiting Vallenar expressed similar sentiments. Barrick's programs and the effects on the community are depicted in the documentary *El Tesoro de América*. See also Campisi 2008; Li 2017; Urkidi 2010.

29. State official 4, interview by author, July 2011.

30. State official 4, interview by author, July 2011.

31. Using data from Pascua Lama's glacier monitoring plan, in 2011 scientists estimated that the three glaciers contribute up to 22 percent of the water at the top of the Toro River in summer (Gascoin et al. 2011)—a much larger amount than scientists claimed in the EIA studies. Gascoin and his coauthors do not say whether this was the pre–Pascua Lama background rate, or if it results from accelerated glacier melt.

32. State official 2, interview by author, July 2011; Escobar 2005.

33. Scientists 1, 2, and 3, interviews by author, July 2011. According to Campisi (2008), Barrick promoted climate change as a hypothesis to explain the loss of these glaciers to deflect attention from its own responsibility in destroying them. The company went so far as to invite former US vice president Al Gore, then promoting his film related to climate change, *An Inconvenient Truth*, to Chile. Pressured by activists, Gore finally declined the invitation.

34. Using monitoring data from Pascua Lama, glaciologists have since estimated that the three affected glaciers have lost as much as 79 percent of their surface area in the twentieth century—far more than the 9 percent lost by the nearby Ortigas glacier (Rabatel et al. 2011). Like Simon Gascoin and colleagues (2011), Antoine Rabatel and colleagues avoid discussing what impacts, if any, Barrick's mining activities might have had in the past or present on the glaciers, although they do suggest that dust deposition is a major problem.

35. NGOs opposed to Barrick were disappointed with the verdict but celebrated that the judges indirectly recognized that the glaciers have suffered human-caused damages (Asamblea por el Agua 2015).

36. On the value and challenges of glacier inventories, see Taillant 2015.

37. Scientists 1 and 4, and state official 1, interviews by author, June–July 2011.

38. As of this writing in mid-2017, the mine's operations were still suspended due to environmental infractions, lawsuits (including a class action suit brought by shareholders in the United States), and poor economic results. Barrick was seeking to revitalize the project with a new partner (Shandong Gold) and plans to build an underground rather than open-pit mine.

Chapter 6

1. Patricio Rodrigo, director of Patagonia without Dams, raised all these issues in a short interview with reporter Daniela Ceballos on CNN Chile: his group opposed HidroAysén because it centralized political and economic power to benefit a few private business interests. Aired on August, 22, 2009, accessed November 5, 2017, https://www.youtube.com/watch?v=F-CVnjPrzlA.

2. The electricity was going to be transported by a two-thousand-kilometers-long high-voltage transmission line to central Chile. The transmission line was evaluated in a separate EIA that was never finished due to ongoing controversy.

3. Data from an IPSOS survey (Liberona 2011). On May 15, 2011, *La Tercera* newspaper reported that 74 percent rejected the project. I could not find opinion data that are representative of the population of Aysén.

4. This decision was confirmed one year later by the energy ministry, revoking HidroAysén's EIA permit. In parallel, the project faced at least nine legal challenges in Chile's court system (Eyzaguirre and Sottovia 2014).

5. Elites of all political parties share this concern. Socialist president Lagos expressed similar ideas during the Celco Arauco conflict (chapter 4; Sepúlveda 2016), and conservative Energy Minister Laurence Golborne (2011) echoed this fear in the run-up to HidroAysén's first approval and again after the project's demise (Laurence 2014).

6. Scientist 21, interview by author, January 2011. Scientists 11, 13, 15, and 22, activist 6, and consultants 2 and 11 all complained about the lack of research and monitoring infrastructure in Aysén. These kinds of complaints were common across all four conflicts covered in this book. But the case of HidroAysén has a twist, as some scientists anxiously felt that Aysén's nature was unknowable, not just unknown. The implication is that the lack of knowledge about the region was so profound that to collect the information needed to properly assess a project like HidroAysén required superhuman efforts.

7. Statement by Pablo Osses (HidroAysén 2008). A decade earlier, scientists from the Austral University who participated in the EIA studies for the Valdivia paper and pulp mill expressed similar motivations for wanting to participate in those studies (Sepúlveda 2016).

8. Author's calculation. The 2008 budget for Fondecyt was about 45 billion Chilean pesos (Conicyt website), which is about US$86 million (at the 2008 exchange rate). Fondecyt is the most important source of funding for Chilean scientists (Barandiarán 2012). Scientists 11, 12, 16, 17, and 25 emphasized the opportunities and money that HidroAysén's EIA provided.

9. Industry staff 9, interview by author, May 2011; see also statements by Salazar quoted in Pizarro 2014.

10. Activist 5, and state officials 10 and 13, interviews by author, March and June 2011.

11. What made the original maps Endesa submitted illegible is that they were reproduced on A4 pages, in pdf documents, at scales of 1:100.000, 1:150.000, or 1:250.000. The text on the labels and geographic points of reference was often too small to be legible. I have not reproduced a map here because, in black and white at this size, a sample would have looked even worse than the originals.

12. Industry staff 8, interview by author, March 2011.

13. Consultants 4 and 12, and industry staff 9, interviews by author, April–May 2011.

14. Conama's National Information Service was never able to create such a database. Under the new Environment Ministry, the National Information Service improved.

Like the proposals to integrate environmental data regarding aquaculture discussed in chapter 3, the service seeks to "capture/integrate/generate and provide access to environmental information" (http://sinia.mma.gob.cl/componentes-del-sinia/). As of July 2017, the search engine included several interesting publications, but the website did not specify which, if any of these, had been produced for an EIA. In parallel, the EIA Agency maintains a geographic database of EIA studies (http://sig .sea.gob.cl/mapaLineasBaseEIA/).

15. Activist 7, interview by author, March 2011.

16. Scientist 22, interview by author, July 2011.

17. Activist 5, interview by author, March 2011.

18. Consultants 1, 2, 3, 4, and 12, interviews by author, December 2010–April 2011.

19. Scientist 13, interview by author, February 2011.

20. Scientist 12, interview by author, March 2011.

21. Consultants 1, 11, and 12, interviews by author, December 2010–April 2011.

22. Scientist 18, interview by author, November 2010.

23. Scientists 6, 11, 12, 14, 21, 25, and 27, state officials 10, 13, 17, and 28, and activists 6, 7, and 12, interviews by author, November 2010–April 2011. Many interviewees were convinced that EIA science is poor quality because of conflicts of interest.

24. Scientist 18, interview by author, November 2010. Scientists 11, 12, 14, 16, and 17 made similar remarks.

25. Scientist 17, interview by author, May 2011.

26. Industry staff 9, interview by author, May 2011. Industry staff 8 and state official 15 expressed similar ideals. In Peru, actors use a similar distinction between baseline studies and impact evaluations (Li 2015).

27. Complaints of cut and paste along with otherwise poor-quality EIAs were common (e.g., scientists 11, 18, 21, and 25, and state officials 10, 17, and 28).

28. Scientist 17, interview by author, May 2011.

29. Scientists 11, 12, 13, 14, 16, 17, 18, 21, and 22 described these dynamics. A decade earlier, Austral scientists who worked on Celco Arauco's EIA studies expressed similar sentiments and complaints (Sepúlveda 2016).

30. Scientist 21 and industry staff 8, interviews by author, January and March 2011.

31. Consultant 2, interview by author, December 2010.

32. Industry staff 9, interview by author, May 2011.

33. Scientists 13, 14, 16, 21, 24, and 25, interviews by author, December 2010–July 2011.

34. Industry staff 8, interview by author, March 2011. Endesa was also interested in protecting data it considered its private property. The company acquiesced to some demands from the Water Agency, explaining that it was releasing the data "only to demonstrate that we proceed with the utmost caution in interpreting the data" (Endesa 2010, 772).

35. Two other such centers, CEAZA and COPAS, played important roles in the other conflicts analyzed in this book. See also Baigorrotegui and Santander 2018.

36. Scientist 12, interview by author, March 2011.

37. Consultant 3, interview by author, April 2011.

38. Scientist 11, interview by author, March 2011.

39. This fear was widespread (as in statements by scientists 16, 17, 18, and 22), but also countered by others (e.g., scientists 11, 12, 13, and 21, along with activists 5, 6, and 7, and consultant 12) who had harsh words for scientists who kept their misgivings silent to protect their fragile reputations as supposedly objective.

40. Industry staff 7, interview by author, February 2011.

41. State official 16, interview by author, March 2011.

42. Moreover, many complained that poor-quality baselines obscured a project's impacts (consultant 2, state officials 10, 13, 14, and 17, and activists 5 and 6) or that officials simply pay too much attention to baselines when impact evaluations are more important (consultant 12 and state official 12).

43. Consultant 2, interview by author, December 2010.

44. Industry staff 6, interview by author, March 2011.

45. State official 20, interview by author, June 2011.

46. TVN-24horas, April 18, 2011, accessed May 15, 2011, http://www.24horas.cl/videos.aspx?id=115236&tipo=410; TeleSur, May 18, 2011, accessed November 8, 2017, https://youtu.be/DXuqVtBqc_8; CNN Chile, August, 22, 2009, accessed November 8, 2017, https://www.youtube.com/watch?v=F-CVnjPrzlA).

47. I assume the comments are from Chilean scientists; the names are in Spanish, the comments indicate intimate knowledge of Chilean environmental politics, and only universities have access to *Nature* in Chile. See Gardner 2011.

48. State official 14, interview by author, March 2011.

49. State official 15, interview by author, March 2011.

50. State officials 12, 14, 15, and 27, interviews by author, November 2010–March 2011.

51. The reasons given to cancel HidroAysén in 2014 could have been used to deny its original EIA permit. See Buschheister 2014; Gobierno rechaza 2014.

52. That Obama's successor overturned this decision only highlights how political EIA decision-making can be.

Conclusion

1. For a detailed discussion of the permutations and breakdowns of this ideal during the twentieth century, and its continued relevance to liberal democracy, see Ezrahi 1990.

2. The Environmental Data and Governance Initiative has documented these and other changes to the EPA introduced by the Trump administration. See https://100 days.envirodatagov.org/, accessed March 10, 2018.

3. See http://www.marchforscience.com/, accessed March 10, 2018.

4. The Oxford Dictionary made post-truth the 2016 "word of the year." See https:// en.oxforddictionaries.com/word-of-the-year/word-of-the-year-2016, accessed March 10, 2018. Online articles on the topic appear in the *New Statesman* and on blogs by British universities and Oxford University Press. British commentators have reflected on what post-truth politics means for universities, civil servants, and math teachers, to name a few topics covered online.

5. By contrast, stronger environmental quality standards and emission restrictions would improve environmental outcomes.

6. For a parallel experience in the United States, see Lave 2012.

7. Gwen Ottinger (2013) finds similar problems in her study of toxic pollution in the United States.

References

Actores relevantes se refieren a renuncia del director del servicio de evaluación ambiental. 2011. *Diario de Aysén*, April 29.

Adas, Michael. 2014. *Machines as the Measure of Men: Science, Technology, and Ideologies of Western Dominance*. Ithaca, NY: Cornell University Press.

Aguila, Ernesto. 2011. Deficit Democrático. *La Tercera*, May 11.

Ahmad, Balsam, and Christopher Wood. 2002. A Comparative Evaluation of the EIA Systems in Egypt, Turkey, and Tunisia. *EIA Review* 22 (3): 213–234.

A la vanguardia en la conservación de nuestra biodiversidad. N.d. Catholic University of Chile. Accessed March 21, 2016, http://investigacion.uc.cl/Investigacion-en -la-UC/a-la-vanguardia-en-la-conservacion-de-nuestra-biodiversidad.html.

Allende, Jorge, Jorge Babul, Servet Martinez, and Tito Ureta. 2005. *Análisis y proyecciones de la ciencia chilena*. Santiago: Academia Chilena de Ciencias, Consejo de Sociedades Científicas de Chile.

Altieri, Miguel, and Alejandro Rojas. 1999. Ecological Impacts of Chile's Neoliberal Policies. *Environment, Development, and Sustainability* 1 (1): 55–72.

Alvarez, Daniel. 2001. Un valle con futuro. *Pastoral Popular*, November–December, n.p.

Alvial, Adolfo, Frederick Kibenge, John Forster, José M. Burgos, Rolando Ibarra, and Sophie St-Hilaire. 2012. *The Recovery of the Chilean Salmon Industry: The ISA Crisis and Its Consequences and Lessons*. Puerto Montt: World Bank Group.

Ambientalistas navegaron el Mapocho en plena inaguración del museo de la luz. 2011. *Tele 13*, January 19.

Andermann, Jens. 2007. *The Optic of the State: Visuality and Power in Argentina and Brazil*. Pittsburgh: University of Pittsburgh Press.

Appelbaum, Nancy. 2013. Envisioning the Nation: The Mid-Nineteenth Century Colombian Cartographic Commission. In *State and Nation Making in Latin America and Spain: Republics of the Possible*, ed. Miguel Centeno and Agustin Ferraro. Cambridge: University of Cambridge Press.

Appiah-Opoku, Seth. 2005. *The Need for Indigenous Knowledge in Environmental Impact Assessment: The Case of Ghana*. Lewiston, NY: Edwin Mellen Press.

Arenas, Federico. 1988. Medio ambiente: Un nuevo contexto para un nuevo encuentro. *Ambiente y Desarrollo* 4 (3): 77–86.

Armesto, Juan, Ricardo Rozzi, Pablo Marquet, Olga Barbosa, Aurora Gaxiola, Francisca Massardo, Juan Luis Celis, et al. 2011. Letter to the Editor. *La Tercera*, May 28.

Artacho, Paulina, Mauricio Soto-Gamboa, Claudio Verdugo, and Roberto F. Nespolo. 2007a. Blood Biochemistry Reveals Malnutrition in Black-Necked Swans (Cygnus Melanocoryphus) Living in a Conservation Priority Area. *Comparative Biochemistry and Physiology Part A: Molecular & Integrative Physiology* 146 (2): 283–290.

Artacho, Paulina, Mauricio Soto-Gamboa, Claudio Verdugo, and Roberto F. Nespolo. 2007b. Using Haematological Parameters to Infer the Health and Nutritional Status of an Endangered Black-Necked Swan Population. *Comparative Biochemistry and Physiology Part A: Molecular & Integrative Physiology* 147 (4): 1060–1066.

Asamblea por el Agua en el Alto Huasco. 2015. Fallo del Tribunal ambiental reconoce que existían pruebas suficientes de destrucción de glaciares pero no condenaron a Barrick Gold. OLCA Comunicaciones, March 24.

Asche, Frank, Havard Hansen, Ragnar Tveterassigbjorn, and Sigbjorn Tveteras. 2010. The Salmon Disease Crisis in Chile. *Marine Resource Economics* 25:405–411.

Astudillo, Antonio, and Constanza Pérez-Cueto. 2014. Endesa ve cambio de actitud del gobierno hacia HidroAysén. *La Tercera*, June 12.

Austral. 2005. *Informe Final: Estudio sobre origen de mortalidades y disminución poblacional de aves acuáticas en el Santuario de la Naturaleza Carlos Anwandter, en la provincia de Valdivia*. Prepared by Austral University for Dirección Regional de Conama, X Región de Los Lagos. Accessed October 30, 2017, http://www.accionporloscisnes .org/documentos/seia/informe_%20uach_final.pdf.

Baigorrotegui, Gloria, and María Teresa Santander. 2018. Localities Facing the Construction of Fossil Power Plants: Two Experiences to Discuss the Hostile Side of Electricity Infrastructures. In *Spanish Philosophy of Technology: Contemporary Work from the Spanish Speaking Community*, ed. Bélen Laspra and José Antonio López Cerezo. New York: Springer.

Barandiarán, Javiera. 2012. Threats and Opportunities of Science at a For-Profit University in Chile. *Higher Education* 63 (2): 205–218.

Barandiarán, Javiera. 2015a. Chile's Environmental Assessments: Contested Knowledge in an Emerging Democracy. *Science as Culture* 24 (3): 251–275.

Barandiarán, Javiera. 2015b. Reaching for the Stars? Astronomy and Growth in Chile. *Minerva* 53 (2): 141–164.

Barandiarán, Javiera. 2016. The Authority of Rules in Chile's Contentious Environmental Politics. *Environmental Politics* 25 (6): 1013–1033.

Barcaza, Gonzalo, and Pablo Wainstein. 2002. Implicancias ambientales del Proyecto Pascua Lama. *Pastoral Popular*, May–June, n.p.

Barrick. 2000. *Estudio de impacto ambiental: Pascua Lama.* Accessed November 2, 2017, http://www.sea.gob.cl/.

Barrick. 2004. *Estudio impacto ambiental: Modificaciones al proyecto Pascua Lama.* Prepared by Arcadis. Accessed November 2, 2017, http://www.sea.gob.cl/.

Barrick. 2005a. *Adenda No. 1: Estudio de impacto ambiental "modificaciones al proyecto Pascua Lama."* Prepared by Arcadis. Accessed November 2, 2017, http://www.sea .gob.cl/.

Barrick. 2005b. *Adenda No. 2: Estudio de impacto ambiental "modificaciones al proyecto Pascua Lama."* Prepared by Arcadis. Accessed November 2, 2017, http://www.sea .gob.cl/.

Barrick. 2006. *Adenda No. 3: Estudio de impacto ambiental "modificaciones al proyecto Pascua Lama."* Prepared by Arcadis. Accessed November 2, 2017, http://www.sea .gob.cl/.

Barton, Jonathan, and Arnt Floysand. 2010. The Political Ecology of Chilean Salmon Aquaculture, 1982–2010: A Trajectory from Economic Development to Global Sustainability. *Global Environmental Change* 20 (4): 739–752.

Bauer, Carl. 1998. *Against the Current: Privatization, Water Markets, and the State in Chile.* Boston: Kluwer Academic Publishers.

Beigel, Fernanda, ed. 2013. *The Politics of Academic Autonomy in Latin America.* Surrey, UK: Ashgate Press.

Benchimol, Jaime. 1999. *Dos micróbios aos mosquitos: Febre amarela e a revoluçao pasteriana no Brasil.* Rio de Janeiro: Editora Fiocruz y Editora UFRJ.

Benítez, Andrés. 2010. ¿The Chilean Way? *La Tercera*, October 23. Accessed October 22, 2017, http://www.latercera.com/noticia/opinion/ideas-y-debates/2010/10/895 -301780-9-the-chilean-way.shtml.

Berman, Elizabeth Popp. 2012. *Creating the Market University: How Academic Science Became an Economic Engine.* Princeton, NJ: Princeton University Press.

Bernasconi, Andrés. 2008. Chile. In *Universidad y desarrollo en Latinoamérica: Experiencias exitosas de centros de investigación,* ed. Simon Schwartzman, 211–240. Bogotá: IESALC-UNESCO.

Bhadra, Monamie. 2013. Fighting Nuclear Energy, Fighting for India's Democracy. *Science as Culture* 22 (2): 238–246.

Bijker, Wiebe E., Roland Bal, and Ruud Hendriks. 2009. *The Paradox of Scientific Authority: The Role of Scientific Advice in Democracies.* Cambridge, MA: MIT Press.

Bojórquez-Tapia, Luis, and Ofelia García. 1998. An Approach for Evaluating EIAs: Deficiencies of EIA in Mexico. *EIA Review* 18 (3): 217–240.

Bórquez, Roxana. 2007. Análisis del escenario actual de los glaciares de montaña en Chile desde la mirada de la seguridad ecológica. MA thesis, University of Chile.

Bravo, Sanda, María Teresa Silva, and Claudia Lagos. 2007. *Diagnóstico de la proyección de la investigación en ciencia y tecnología de la acuicultura chilena.* Austral University of Chile. On file with author.

Brenning, Alexander. 2008. The Impact of Mining on Rock Glaciers. In *Darkening Peaks: Glacier Retreat, Science, and Society*, ed. Ben Orlove, Ellen Wiegandt, and Brian H. Luckman, 196–205. Berkeley: University of California Press.

Brown, Mark B. 2009. *Science in Democracy: Expertise, Institutions, and Representations.* Cambridge, MA: MIT Press.

Brunner, José Joaquín. 1986. *Informe Sobre la Educacion Superior en Chile.* Santiago: FLACSO.

Brunner, José Joaquín. 1993. Chile's Higher Education: Between Market and State. *Higher Education* 25 (1): 35–43.

Buch, Alfonso. 2006. *Forma y función de un sujeto moderno: Bernardo Houssay y la fisiología Argentina.* Buenos Aires: Universidad Nacional de Quilmes Editorial.

Burnett, Graham. 2000. *Masters of All They Surveyed: Explorations, Goegraphy, and a British El Dorado.* Chicago: University of Chicago Press.

Bury, Jeffrey. 2015. The Frozen Frontier: The Extractives Super Cycle in a Time of Glacier Recession. In *The High-Mountain Cryosphere: Environmental Changes and Human Risks*, ed. Christian Huggel, John Clague, Andreas Kaab, and Mark Carey, 71–89. Cambridge: Cambridge University Press.

Buschheister, Axel. 2014. Oportunismo y elusión. *La Tercera*, June 15.

Buschmann, Alejandro, Verónica Riquelme, María Hernández-González, Daniel Varela, Jaime Jiménez, Luis A. Henríquez, Pedro A. Vergara, Ricardo Guíñez, and Luis Filún. 2006. A Review of the Impacts of Salmonid Farming on Marine Coastal Ecosystems in the Southeast Pacific. *ICES Journal of Marine Science* 63 (7): 1338–1345.

Bustos, Beatriz. 2015. Producción de conocimiento en Chile y el caso de la crisis salmonera. In *Ecología política en Chile: Naturaleza, propiedad, conocimiento y poder*, ed. Beatriz Bustos, Manuel Prieto, and Jonathan Barton, 193–212. Santiago: Editorial Universitaria.

Bustos, Beatriz, Manuel Prieto, and Jonathan Barton, eds. 2015. *Ecología política en Chile: Naturaleza, propiedad, conocimiento y poder*. Santiago: Editorial Universitaria.

Caldwell, Lynton K. 1982. *Science and the National Environmental Policy Act: Redirecting Policy through Procedural Reform*. Birmingham: University of Alabama Press.

Callon, Michel, Pierre Lascoumes, and Yannick Barthe. 2012. *Acting in an Uncertain World: An Essay on Technical Democracy*. Cambridge, MA: MIT Press.

Campisi, Christina. 2008. Constituting "Community" at the Onset of the Pascua Lama Mining Project. MA thesis, University of Montreal.

Camus, Pablo, and Ernst Hajek. 1998. *Historia ambiental de Chile*. Santiago: Andros Impresores.

Carey, Mark. 2010. *In the Shadow of Melting Glaciers: Climate Change and Andean Society*. Oxford: Oxford University Press.

Carmona, Alejandra. 2011. Ex seremi de Vivienda y parlamentarios cuestionan procedimiento de evaluación del megaproyecto. *El Mostrador*, May 5.

Carruthers, David. 2001. Environmental Politics in Chile: Legacies of Dictatorship and Democracy. *Third World Quarterly* 22 (3): 343–358.

Carruthers, David, ed. 2008. *Environmental Justice in Latin America: Problems, Promise, and Practice*. Cambridge, MA: MIT Press.

Carruthers, David, and Patricia Rodriguez. 2009. Mapuche Protest, Environmental Conflict, and Social Movement Linkage in Chile. *Third World Quarterly* 30 (4): 743–760.

CASEB. 2005. *Comentarios sobre el informe final de la Universidad Austral de Chile para la Dirección Regional de Conama, X Región de Los Lagos, Estudio sobre origen de mortalidades y disminución poblacional de aves acuáticas en el Santuario de la Naturaleza Carlos Anwandter, en la provincia de Valdivia*. Prepared by the Centro de Estudios Avanzados en Ecología y Biodiversidad (CASEB) of the Pontificia Universidad Católica de Chile. On file with author.

Cashmore, Matthew, and Tim Richardson. 2013. Power and Environmental Assessment: Introduction to the Special Issue. *EIA Review* 39 (1): 1–4.

Cashmore, Matthew, Alan Bond, and Dick Cobb. 2007. The Contribution of Environmental Assessment to Sustainable Development: Toward a Richer Empirical Understanding. *Environmental Management* 40 (3): 516–530.

CDE (Estado Fisco de Chile) v. Arauco (Celulosa Arauco y Constitución). 2013. Rol. 746–2005, First Civil Court (Valdivia).

Celedón, S., Verónica Moya, and J. Esturillo. 2011. Privados y autoridades critican acción contra funcionarios que evaluarán HidroAysén. *El Mercurio*, April 22.

CENMA. 1999. *Annual Report*. Accessed in the Environment Ministry library.

CENMA. 2008. *Convenio entre la Comisión Nacional del Medio Ambiente y la Fundación Centro Nacional del Medio Ambiente*. On file with author.

CENMA. 2009. *Convenio entre la Comisión Nacional del Medio Ambiente y la Fundación Centro Nacional del Medio Ambiente*. On file with author.

CENMA. 2010. *Convenio entre la Comisión Nacional del Medio Ambiente y la Fundación Centro Nacional del Medio Ambiente*. On file with author.

CENMA. 2014. *Curriculum Institucional del Centro Nacional de Medio Ambiente, July*. On file with author.

Centellas, Kate. 2010. The Localism of Bolivian Science: Tradition, Policy, and Projects. *Latin American Perspectives* 37 (3): 160–175.

Centellas, Kate. 2014. "Cameroon Is Just Like Bolivia!": Southern Expertise and the Construction of Equivalency in South-South Scientific Collaborations. *Information and Culture* 49 (2): 177–203.

Chile Signs Up as First OECD Member in South America. 2010. Organization for Economic Cooperation and Development, November 1. Accessed October 22, 2017, http://www.oecd.org/chile/chilesignsupasfirstoecdmemberinsouthamerica.htm.

Clausen, Alison, Hoang Hoa Vu, and Miguel Pedrono. 2010. An Evaluation of the Environmental Impact Assessment System in Vietnam: The Gap between Theory and Practice. *EIA Review* 31 (2): 136–143.

Collins, Harry M., and Robert Evans. 2002. The Third Wave of Science Studies: Studies of Expertise and Experience. *Social Studies of Science* 32 (235): 235–296.

Conama. 1998. Resolución exenta 279. EIA permit. Accessed October 30, 2017, http://www.sea.gob.cl/.

Conama. 1999. Resolución exenta 009. Modifies EIA permit. Accessed October 30, 2017, http://www.sea.gob.cl/.

Conama. 2001. Resolución exenta 39: Califica estudio de impacto ambiental del proyecto Pascua Lama. EIA permit. Accessed November 2, 2017, http://www.sea.gob.cl/.

Conama. 2006. Resolución exenta 24: Califica estudio de impacto ambiental del proyecto Pascua Lama. EIA permit. Accessed November 2, 2017, http://www.sea.gob.cl/.

Constable, Pamela, and Arturo Valenzuela. 1991. *A Nation of Enemies: Chile under Dictatorship*. New York: W. W. Norton.

Cozzens, Susan. 1990. Autonomy and Power in Science. In *Theories of Science in Society*, ed. Susan Cozzens and Thomas F. Gieryn, 164–184. Bloomington: Indiana University Press.

Croxatto, Hector. 1987. Comentario. *Ambiente y Desarrollo* 3 (1–2): 24–25.

Cruz Pérez v. Barrick (Compañía Minera Nevada SpA). 2015. Rol. 2–2013, Second Environmental Tribunal (Santiago).

Cueto, Marcos. 1989. Andean Biology in Peru: Scientific Styles on the Periphery. *Isis* 80 (4): 640–658.

Cueto, Marcos. 1994. *Missionaries of Science: The Rockefeller Foundation and Latin America*. Bloomington: Indiana University Press.

Cueto, Marcos. 1997. Science under Adversity: Latin American Medical Research and American Private Philanthropy, 1920–1960. *Minerva* 35 (3): 233–245.

Cushman, Gregory. 2013. *Guano and the Opening of the Pacific World: A Global Ecological History*. Cambridge: Cambridge University Press.

Dahl, Richard. 2008. Science in the Courtroom: Examining Standards for Litigation-Based Research. *Environmental Health Perspectives* 116 (1): A37.

Deisley, David. 2001. Letter to the Editor. *Pastoral Popular*, November 19, n.p.

De la Maza, Carmen. 2001. NEPA's Influence in Developing Countries: The Chilean Case. *EIA Review* 21 (2): 169–179.

Delgado, Luisa, Víctor Marín, Pamela Bachmann, and Marcela Torres-Gomez. 2009. Conceptual Models for Ecosystem Management through the Participation of Local Social Actors: The Río Cruces Wetland Conflict. *Ecology and Society* 14 (1): 50–71.

Diaz, Elena, Yolanda Texera, and Hebe Vessuri. 1983. *La ciencia periférica: Ciencia y sociedad en Venezuela*. Caracas: Centro de Estudios del Desarrollo.

Director del CIEP se refiere a crítica formulada por participar en seminario de entrega de líneas bases del EIA de megaproyecto hidroeléctrico. 2008. *El Divisadero*, January 16. Accessed November 7, 2017, http://www.eldivisadero.cl/noticia-25593.

Doberstein, Brent. 2004. EIA Models and Capacity Building in Viet Nam: An Analysis of Development Aid Programs. *EIA Review* 24 (3): 283–318.

Dubé, M. G. 2004. Advances in Assessing the Effects of Pulp and Paper Mill Effluents on Aquatic Systems. In *Paper and Pulp Mill Effluent Environmental Fate and Effects*, ed. Dennis Borton, Timothy J. Hall, Robert P. Fisher, and Jill F. Thomas, 397–409. Lancaster, PA: DEStech Publications.

Duit, Andreas, ed. 2014. *State and Environment*. Cambridge, MA: MIT Press.

Durán, Valentina. 2006. Caso Celco: Un fallo controvertido de la excelentisima Corte Suprema. *Revista de Derecho Ambiental* 2:253–275.

EIA Agency. 2011. Resolución exenta 225: Califica ambientalmente proyecto Hidroeléctrico Aysén. EIA permit. Accessed November 5, 2017, http://www.sea.gob.cl/.

El-Fadl, Karma, and Mutasem El-Fadel. 2004. Comparative Assessment of EIA Systems in MENA Countries: Challenges and Prospects. *EIA Review* 24 (6): 553–593.

Endesa. 2008. *Proyecto Hidroeléctrico Aysén*. Prepared by Sweco-POCH-EPS. Accessed November 6, 2017, http://www.sea.gob.cl/.

Endesa. 2010. *Adenda no. 2: Proyecto Hidroeléctrico Aysén*. Accessed November 6, 2017, http://www.sea.gob.cl/.

Environment Ministry. 2016. *Informe final de la Comisión Asesora Presidencial para la evaluación del SEIA*. Accessed August 7, 2017, http://portal.mma.gob.cl/wp-content /doc/35877_Informe-MMAF_FINAL.pdf.

Es delicado y poco prudente que el CIEP aparezca vinculado a la labor de difusión de HidroAysén. 2008. *El Divisadero*, January 15. Accessed November 7, 2017, http:// www.eldivisadero.cl/noticia-25608.

Escaida, José, Eduardo Jaramillo, Carlos Amtmann, and Nelson A. Lagos. 2014. *El humedal del Río Cruces y el cisne de cuello negro*. Valdivia: Ediciones UACh.

Escobar, Fernando. 2005. *Informe de comisión de servicio a la III Región: Visita a Pascua Lama*. January 12. Water Agency's library, Santiago, n.p.

Espinoza, Guillermo. 2007. *Gestión y fundamentos de evaluación de impacto ambiental*. Santiago: Centro de Estudios para el Desarrollo.

Espinoza, Guillermo, and V. Alzina. 2001. *Review of Environmental Impact Assessment in Selected Countries of Latin America and the Caribbean: Methodology, Results, and Trends*. Washington, DC: Inter-American Development Bank.

Estay, Manuel, and Carlos Chávez. 2015. Decisiones de localización y cambios regulatorios: El caso de la acuicultura en Chile. *Latin American Journal of Aquaculture Research* 43 (4): 700–717.

Ex ministra de Medio Ambiente: La mayor gravedad de HidroAysén es que de concretarse, es irreversible. 2011. *El Divisadero*, March 19.

Eyzaguirre, Antonia, and César Sottovia. 2014. Comité de Ministros rechaza en forma unánime proyecto HidroAysén. *El Mercurio*, June 11.

Ezrahi, Yaron. 1990. *The Descent of Icarus: Science and the Transformation of Contemporary Democracy*. Cambridge, MA: Harvard University Press.

Fajardo, Marco. 2015. Biólogo Andrés Couve y la rebelión de los científicos: El problema es que Chile no cree en su cultura. *El Mostrador*, November 12.

Farrell, Alexander E., Stacy VanDeveer, and Jill Jager. 2001. Environmental Assessments: Four Under-Appreciated Elements of Design. *Global Environmental Change* 11 (4): 311–333.

Figueroa, Francisco. 2013. *Llegamos para quedarnos: Crónicas de la revuelta estudiantil.* Santiago: LOM Ediciones.

Fischer, Frank. 2000. *Citizens, Experts, and the Environment: The Politics of Local Knowledge.* Durham, NC: Duke University Press.

Fischer, Karen. 2009. The Influence of Neoliberals in Chile before, during, and after Pinochet. In *The Road from Mont Pelerin,* ed. Philip Mirowski and Dieter Plehwe, 305–346. Cambridge, MA: Harvard University Press.

Fortun, Kim. 2001. *Advocacy after Bhopal: Environmentalism, Disaster, New Global Orders.* Chicago: University of Chicago Press.

Frickel, Scott, and Debra J. Davidson. 2004. Building Environmental States: Legitimacy and Rationalization in Sustainability Governance. *International Sociology* 19 (1): 89–110.

Frickel, Scott, and Kelly Moore, eds. 2005. *The New Political Sociology of Science: Institutions, Networks, Power.* Madison: University of Wisconsin Press.

Friedman, Milton. 1962. *Capitalism and Freedom.* Chicago: University of Chicago Press.

Friedman, Milton. 1982. Free Markets and the Generals. *Newsweek,* January 25.

Fuentes Olmos, Jessica. 2014. Evolución del régimen ambiental de la acuicultura en Chile. *Revista de Derecho de la Pontificia Universidad Católica de Valparaíso* 42:441–477.

Galison, Peter. 1997. *Image and Logic: A Material Culture of Microphysics.* Chicago: University of Chicago Press.

Garcés, Mario. 2012. *El despertar de la sociedad: Los movimientos sociales en América Latina y Chile.* Santiago: LOM Ediciones.

Gardner, Elie. 2011. Massive Chilean Dams Approved. *Nature,* May 17. doi:10.1038/news.2011.297.

Garretón, Miguel A. 2003. *Incomplete Democracy: Political Democratization in Chile and Latin America.* Chapel Hill: University of North Carolina Press.

Gascoin, Simon, Christophe Kinnard, Rodrigo Ponce, Shelley Macdonell, Stefaan Lhermitte, and Antoine Rabatel. 2011. Glacier Contribution to Streamflow in Two Headwaters of the Huasco River, Dry Andes of Chile. *Cryosphere* 5 (4): 1099–1113.

Geiger, Roger. 2004. *Knowledge and Money: Research Universities and the Paradox of the Marketplace.* Palo Alto, CA: Stanford University Press.

Gerhart, Andrew. 2017. Petri Dishes of an Archipelago: The Ecological Rubble of the Chilean Salmon Farming Industry. *Journal of Political Ecology* 24 (41): 726–742.

Gieryn, Thomas F. 1999. *Cultural Boundaries of Science: Credibility on the Line.* Chicago: University of Chicago Press.

Girardi, Guido, and Juan A. Asenjo. 2011. Los científicos en la información de la nación. *El Mercurio*, April 29.

Glasson, John, and Nemesio Neves Salvador. 2000. EIA in Brazil: A Procedures-Practice Gap. A Comparative Study with Reference to the European Union, and Especially the UK. *EIA Review* 20 (2): 191–225.

Glasson, John, Riki Therivel, and Andrew Chadwick. 2012. *Introduction to Environmental Impact Assessment*. London: Routledge.

Gligo, Nicolo. 2011. Desenmarañando los conflictos de Hidroaysén. *Universidad de Chile* (blog), June 8. Accessed November 5, 2017, http://www.uchile.cl/noticias/72365/nicolo-gligo-desenmaranando-los-conflictos-de-hidroaysen.

Gobierno rechaza proyecto HidroAysén tras seis años de tramitación ambiental. 2014. *La Tercera*, Junio 11.

Godoy, Marcos G., Alejandra Aedo, Molly J. T. Kibenge, David B. Groman, Carmencita V. Yason, Horst Grothusen, Angelica Lisperguer, et al. 2008. First Detection, Isolation, and Molecular Characterization of Infectious Salmon Anaemia Virus Associated with Clinical Disease in Farmed Atlantic Salmon (Salmo Salar) in Chile. *BMC Veterinary Research* 4 (1): 28–40.

Golborne encabeza ofensiva de gobierno para defender aprobación de megacentral. 2011. *La Tercera*, May 11.

Goldman, Michael. 2005. *Imperial Nature*. New Haven, CT: Yale University Press.

González Leiva, José Ignacio. 2007. Primeros levantamientos cartográficos generales de Chile con base científica: Los mapas de Claudio Gay y Amado Pissis. *Revista de Geografía Norte Grande* 38:21–44.

Greenberg, Michael R. 2012. *The Environmental Impact Statement after Two Generations: Managing Environmental Power*. New York: Routledge.

Grindle, Merilee S, ed. 1997. *Getting Good Government: Capacity Building in the Public Sectors of Developing Countries*. Cambridge, MA: Harvard University Press.

Guston, David. 2000. *Between Politics and Science: Assuring the Integrity and Productivity of Research*. Cambridge: Cambridge University Press.

Gyhra, Alberto. 1989. Universidad, medio ambiente y desarrollo. *Ambiente y Desarrollo* 5 (1): 79–83.

Hacking, Ian. 1990. *The Taming of Chance*. Cambridge: Cambridge University Press.

Hajek, Ernst. 1987. Medio ambiente, desarrollo y academia. *Ambiente y Desarrollo* 3 (1–2): 11–16.

Halliburton, D., and L. Maddison. 2004. Overview of Improvements in Effluent Quality as a Result of Changes to the Federal and Provincial Pulp and Paper Mill

Effluent Control Regimes. In *Paper and Pulp Mill Effluent Environmental Fate and Effects*, ed. Dennis Borton, Timothy J. Hall, Robert P. Fisher, and Jill F. Thomas, 534–542. Lancaster, PA: DEStech Publications.

Halwart, Matthias, Doris Soto, and J. Richard Arthur. 2007. *Cage Aquaculture: Regional Reviews and Global Overview*. Rome: Food and Agriculture Organization of the United Nations.

Hayek, Friedrich. 1945. The Use of Knowledge in Society. *American Economic Review* 35 (September): 519–530.

Hays, Samuel. 1959. *Conservation and the Gospel of Efficiency*. Cambridge, MA: Harvard University Press.

Hecht, Gabrielle, ed. 2011. *Entangled Geographies: Empire and Technopolitics in the Global Cold War*. Cambridge, MA: MIT Press.

Herrera, Amílcar. 1974. *Ciencia y política en América Latina*. Buenos Aires: Siglo XXI Editores.

HidroAysén da a conocer información medioambiental de su megaproyecto. 2008. *Diario de Aysén*, January 14.

Hilbink, Lisa. 2007. *Judges beyond Politics in Democracy and Dictatorship: Lessons from Chile*. Cambridge: Cambridge University Press.

Hilgartner, Stephen. 2000. *Science on Stage: Expert Advice as Public Drama*. Palo Alto, CA: Stanford University Press.

Hironaka, Ann. 2002. The Globalization of Environmental Protection: The Case of Environmental Impact Assessment. *International Journal of Comparative Sociology* 43 (1): 65–78.

Hochstetler, Kathryn. 2011. The Politics of Environmental Licensing: Energy Projects of the Past and Future in Brazil. *Studies in Comparative International Development* 46 (4): 349–371.

Hochstetler, Kathryn, and Margaret E. Keck. 2007. *Greening Brazil: Environmental Activism in State and Society*. Durham, NC: Duke University Press.

Hodge, Joseph Morgan. 2007. *The Triumph of the Expert: Agrarian Doctrines of Development and the Legacies of British Colonialism*. Athens: Ohio University Press.

Holder, Jane. 2004. *Environmental Assessment: The Regulation of Decision-Making*. Oxford: Oxford University Press.

Holm, Marius, and Marius Dalen. 2003. *The Environmental Status of Norwegian Aquaculture*. Oslo: Bellona Foundation.

Huggel, Christian, Mark Carey, John Clague, and Andreas Kaab. 2015. Introduction to *The High-Mountain Cryosphere: Environmental Changes and Human Risks*, ed. Christian

Huggel, John Clague, Andreas Kaab, and Mark Carey, 1–6. Cambridge: Cambridge University Press.

Iles, Alastair. 2004. Patching Local and Global Knowledge Together: Citizens Inside the US Chemical Industry. In *Earthly Politics: Local and Global in Environmental Governance*, ed. Sheila Jasanoff and Marybeth Long Martello, 285–308. Cambridge, MA: MIT Press.

Illanes, María Angélica. 1993. *En el nombre del pueblo, del estado y de la ciencia: Historia social de la salud pública, Chile, 1880–1973*. Santiago: Colectivo de Atención Primaria.

Informe sobre el simposio: Rol de los ecólogos en la gestión ambiental a raíz del caso del Río Cruces. 2005. Report, November 16. Conference, Pucón, Chile, October 13–16.

Ivanoff, Danka. 2007. *Lago General Carrera: Temporales de Sueños*. Santiago: LOM.

Jaksic, Fabián, Pablo Camus, and Sergio A. Castro. 2012. *Ecología y ciencias naturales: Historia del conocimiento del patrimonio biológico de Chile*. Santiago: Dirección de Bibliotecas.

Jaksic, Iván. 1989. *Academic Rebels in Chile: The Role of Philosophy in Higher Education and Politics*. Albany: State University of New York Press.

Jaksic, Iván. 2013. Ideological Pragmatism and Nonpartisan Expertise in 19th Century Chile: Andres Bello's Contribution to State and Nation Building. In *State and Nation Making in Latin America and Spain*, ed. Miguel Centeno and Agustin Ferraro, 183–202. Cambridge: Cambridge University Press.

Jaramillo, Eduardo, Roberto Schlatter, Heraldo Contreras, Cristian Duarte, Nelson A. Lagos, Enrique Paredes, Jorge Ulloa, Gaston Valenzuela, Bruno Peruzzo, and Ricardo Silva. 2007. Emigration and Mortality of Black-Necked Swans (Cygnus Melancoryphus) and Disappearance of the Macrophyte Egeria Densa in a Ramsar Wetland Site of Southern Chile. *Ambio* 36 (7): 607–609.

Jasanoff, Sheila. 1990. *The Fifth Branch: Science Advisers as Policymakers*. Cambridge, MA: Harvard University Press.

Jasanoff, Sheila. 1995. *Science at the Bar: Law, Science, and Technology in America*. Cambridge, MA: Harvard University Press.

Jasanoff, Sheila. 2004. Ordering Knowledge, Ordering Society. In *States of Knowledge: The Co-Production of Science and Social Order*, ed. Sheila Jasanoff, 13–45. London: Routledge.

Jasanoff, Sheila. 2005. *Designs on Nature: Science and Democracy in Europe and the United States*. Princeton, NJ: Princeton University Press.

Jasanoff, Sheila, and Marybeth Martello. 2004. Conclusion: Knowledge and Governance. In *Earthly Politics: Local and Global in Environmental Governance*, ed. Sheila Jasanoff and Marybeth Long Martello, 335–350. Cambridge, MA: MIT Press.

Jarur, Paola, and Germán Maldonado. 2005. Pascua Lama anuncia: Estamos dispuestos a modificar el proyecto. *El Mercurio*, July 6.

JICA. 2002. *Terminal Evaluation: The National Center for the Environment Project.* Tokyo: Japan International Cooperation Agency.

JICA. 2003. *Environmental Center Approach: Development of Social Capacity for Environmental Management in Developing Countries and Japan's Environmental Cooperation.* Accessed July 25, 2016, http://www.jica.go.jp/english/our_work/evaluation/tech_and _grant/program/thematic/200303_2/.

Junta. 2005. *Protocolo entre Junta de Vigilancia de la cuenca del Río Huasco y sus afluentes y Compañía Minera Nevada, Ltda. [Barrick].* Signed June 30. On file with author.

Junta. 2006. Memoria del Directorio 2004–2006. Prepared by Junta de Vigilancia de la Cuenca del Río Huasco y sus Afluentes. Accessed July 6, 2011, www.riohuasco.cl.

Kalin Arroyo, Mary, Claudio Donoso, Roberto Murúa, Edmundo Pisano, Ricardo Schlatter, and Italo Serey. 1996. *Toward an Ecologically Sustainable Forestry Project: Concepts, Analysis, and Recommendations: Protecting Biodiversity and Ecosystem Processes in the Río Cóndor Project—Tierra Del Fuego.* Santiago: University of Chile.

Keller, Ann Campbell. 2009. *Science in Environmental Policy: The Politics of Objective Advice.* Cambridge, MA: MIT Press.

Khagram, Sanjeev. 2004. *Dams and Development: Transnational Struggles for Water and Power.* Ithaca, NY: Cornell University Press.

Khanna, Tarun, and Yishay Yafeh. 2007. Business Groups in Emerging Markets: Paragons or Parasites? *Journal of Economic Literature* 45 (June): 331–372.

Kinchy, Abby. 2012. *Seeds, Science, and Struggle: The Global Politics of Transgenic Crops.* Cambridge, MA: MIT Press.

Kinchy, Abby, and Daniel Kleinman. 2003. Organizing Credibility: Discursive and Organizational Orthodoxy on the Borders of Ecology and Politics. *Social Studies of Science* 33 (6): 869–896.

Kirp, David. 2003. *Shakespeare, Einstein, and the Bottom Line.* Cambridge, MA: Harvard University Press.

Klubock, Thomas M. 2014. *La Frontera: Forests and Ecological Conflict in Chile's Frontier Territory.* Durham, NC: Duke University Press.

Kolhoff, Arend J, Peter P. J. Driessen, and Hens A. C. Runhaar. 2013. An Analysis Framework for Characterizing and Explaining Development of EIA Legislation in Developing Countries—Illustrated for Georgia, Ghana, and Yemen. *EIA Review* 38 (January): 1–15.

Kreimer, Pablo. 2006. ¿Dependientes o integrados? La ciencia latinoamericana y la nueva división internacional del trabajo. *Nómadas* (April): 199–212.

Kreimer, Pablo, and Manuel Lugones. 2002. Rowing against the Tide: Emergence and Consolidation of Molecular Biology in Argentina, 1960–1990. *Science, Technology, and Society* 7 (2): 285–311.

Krige, John. 2008. *American Hegemony and the Postwar Reconstruction of Science in Europe.* Cambridge, MA: MIT Press.

Krimsky, Sheldon. 2003. *Science in the Private Interest: Has the Lure of Profits Corrupted Biomedical Research?* Lanham, MD: Rowman and Littlefield Publishers.

Kronenberg, Jakub. 2013. Linking Ecological Economics and Political Ecology to Study Mining, Glaciers, and Global Warming. *Environmental Policy and Governance* 23 (2): 75–90.

Lachmund, Jens. 2013. *Greening Berlin: The Co-Production of Science, Politics, and Urban Nature.* Cambridge, MA: MIT Press.

Lagos, Nelson A., Pedro Paolini, Eduardo Jaramillo, Charlotte Lovengreen, Cristian Duarte, and Heraldo Contreras. 2008. Environmental Processes, Water Quality Degradation, and Decline of Waterbird Populations in the Rio Cruces Wetland, Chile. *Wetlands* 28 (4): 938–950.

Lagos, Ricardo, Blake Hounshell, and Elizabeth Dickinson. 2012. *The Southern Tiger: Chile's Fight for a Democratic and Prosperous Future.* New York: Palgrave Macmillan.

Lahsen, Myanna. 2004. Transnational Locals: Brazilian Experiences of the Climate Regime. In *Earthly Politics: Local and Global in Environmental Governance*, ed. Sheila Jasanoff and Marybeth Long Martello, 151–172. Cambridge, MA: MIT Press.

Larraín, Sara. 1999. *Chile sustentable: Propuesta ciudadana para el cambio.* Santiago: LOM Editores.

Latour, Bruno, and Peter Weibel. 2005. *Making Things Public: Atmospheres of Democracy.* Cambridge, MA: MIT Press.

Laurence Golborne, ex ministro de Energía: Obviamente fue una decisión política. 2014. *El Mercurio*, June 15.

Lave, Rebecca. 2012. *Fields and Streams: Stream Restoration, Neoliberalism, and the Future of Environmental Science.* Athens: University of Georgia Press.

Lave, Rebecca, Philip Mirowski, and Samuel Randalls. 2010. Introduction: STS and Neoliberal Science. *Social Studies of Science* 40 (5): 659–675.

Lee, Norman, and George Clive, eds. 2000. *Environmental Assessments in Developing and Transitional Countries.* Chichester, UK: John Wiley.

Lehtinen, K. J. 2004. Relationship of the Technical Development of Pulping and Bleaching to Effluent Quality and Aquatic Toxicity. In *Paper and Pulp Mill Effluent Environmental Fate and Effects*, ed. Dennis Borton, Timothy J. Hall, Robert P. Fisher, and Jill F. Thomas. Lancaster, PA: DEStech Publications.

Leigh Star, Susan. 2010. This Is Not a Boundary Object: Reflections on the Origin of a Concept. *Science, Technology, and Human Values* 35 (5): 601–617.

Leon-Muñoz, Jorge, David Tecklin, Aldo Farias, and Susan Diaz. 2007. Salmonicultura en los lagos del sur de Chile: Ecorregión Valdiviana. Prepared for World Wildlife Fund. On file with author.

Levy, Daniel C. 1986. Chilean Universities under the Junta—Regime and Policy. *Latin American Research Review* 21 (3): 95–128.

LH (Legislative History) 19.300. 1994. Prepared by Biblioteca del Congreso Nacional. Accessed November 9, 2017, https://www.bcn.cl/historiadelaley.

LH (Legislative History) 20.417. 2010. Prepared by Biblioteca del Congreso Nacional. Accessed November 9, 2017, https://www.bcn.cl/historiadelaley.

LH (Legislative History) 20.434. 2010. Prepared by Biblioteca del Congreso Nacional. Accessed November 9, 2017, https://www.bcn.cl/historiadelaley.

Li, Fabiana. 2015. *Unearthing Conflict: Corporate Mining, Activism, and Expertise in Peru*. Durham, NC: Duke University Press.

Li, Fabiana. 2017. Moving Glaciers: Remaking Nature and Mineral Extraction in Chile. *Latin American Perspectives*, June 28. https://doi.org/10.1177/0094582X17713757.

Liberona, Flavia. 2011. Percepción ciudadana sobre HidroAysén: el fracaso de Daniel Fernández. *El Mostrador*, April 25.

Liboroin, Max. 2013. "Waste as Profit and Alternative Economies." *Discard Studies* (blog), July 9. Accessed October 28, 2017, https://discardstudies.com/2013/07/09/waste-as-profit-alternative-economies/.

Lien, Marianne. 2015. *Becoming Salmon*. Berkeley: University of California Press.

Little, Cedric, Christian Felzenstein, Eli Gimmon, and Pablo Munoz. 2015. The Business Management of the Chilean Salmon Farming Industry. *Marine Policy* 54:108–117.

Liverman, Diana M., and Silvina Vilas. 2006. Neoliberalism and the Environment in Latin America. *Annual Review of Environment and Resources* 31 (November): 327–363.

López Magnaso, Sebastián. 2012. *Libertad de empresa y no discriminación económica: doctrina y jurisprudencia del Tribunal Constitucional*. Santiago: Cuadernos del Tribunal Constitucional de Chile.

Lovengreen, Charlotte, John Morrow, Eduardo Jaramillo, Nelson A. Lagos, Heraldo Contreras, and Cristian Duarte. 2008. Incident Ultraviolet Radiation and Disappearance of the Aquatic Macrophyte Egeria Densa in a Ramsar Wetlands Site. *CLEAN—Soil, Air, Water* 36 (10–11): 858–862.

Luna Quevedo, Diego, César Padilla Ormeño, and Julián Alcayaga Olivares. 2004. *El Exilio del cóndor: Hegemonía transnacional en la frontera. El tratado minero entre Chile y Argentina*. Santiago: OLCA.

Mahony, Martin. 2014. The Predictive State: Science, Territory, and the Future of the Indian Climate. *Social Studies of Science* 44 (1): 109–133.

Mansuy, Daniel. 2011. Diálogo de Sordos. *La Tercera*, May 18.

Marara, Madeleine, Nick Okello, Zainab Kuhanwa, Wim Douven, Lindsay Beevers, and Jan Leentvaar. 2011. The Importance of Context in Delivering Effective EIA: Case Studies from East Africa. *EIA Review* 31 (3): 286–296.

Mardones, F. O., A. M. Perez, P. Valdes-Donoso, and T. E. Carpenter. 2011. Farm-Level Reproduction Number During an Epidemic of Infectious Salmon Anemia Virus in Southern Chile in 2007–2009. *Preventive Veterinary Medicine* 102 (3): 175–184.

Marín, Víctor H., Antonio Tironi, Luisa E. Delgado, Manuel Contreras, Fernando Novoa, Marcela Torres-Gómez, René Garreaud, Irma Vila, and Italo Serey. 2009. On the Sudden Disappearance of Egeria Densa from a Ramsar Wetland Site of Southern Chile: A Climatic Event Trigger Model. *Ecological Modelling*, 220:1752–1763.

Martínez, Carlos. 2011. Un rechazo ideológico. *La Tercera*, May 11.

Martínez-Alier, Joan. 2002. *The Environmentalism of the Poor: A Study of Ecological Conflicts and Valuation*. Cheltenham, UK: Edward Elgar.

Mathews, Andrew. 2011. *Instituting Nature: Authority, Expertise, and Power in Mexican Forests*. Cambridge, MA: MIT Press.

McCook, Stuart. 2002. *State's of Nature*. Austin: University of Texas Press.

McElroy, Damien. 2010. Chile's President Brings Mine Rocks for the Queen and David Cameron. *Telegraph*, October 17.

McPhee, James. 2011. Letter to the Editor. *La Tercera*, May 27.

Medina, Eden. 2011. *Cybernetic Revolutionaries: Technology and Politics in Allende's Chile*. Cambridge, MA: MIT Press.

Medina, Eden, Ivan da Costa Marques, and Christina Holmes. 2014. Introduction: Beyond Imported Magic. In *Beyond Imported Magic: Essays on Science, Technology, and Society in Latin America*, ed. Eden Medina, Ivan da Costa Marquez and Christina Holmes, 1–26. Cambridge, MA: MIT Press.

Meier, Claudio. 2011. Hidroelectricidad realmente sustentable para Chile. Distributed by e-mail. On file with author.

Mena, Marcelo, and Pedro Rivera. 2011. Operación HidroAysén. *La Tercera*, May 10.

Milana, Juan Pablo. 2005. *Línea base de la criósfera, proyecto Pascua Lama*. Prepared for the Junta de Vigilancia de Regantes del Río Huasco. On file with author.

Milazzo, Paul Charles. 2006. *Unlikely Environmentalists: Congress and the Clean Water Act, 1945–1972*. Wichita: University Press of Kansas.

Miller, Clark. 1998. Extending Assessment Communities to Developing Countries. ENRP Discussion Paper E-98–15. Kennedy School of Government, Harvard University, 1–61.

Miller, Clark. 2004. Resisting Empire. In *Earthly Politics: Local and Global Environmental Governance*, ed. Sheila Jasanoff and Marybeth Long Martello, 81–102. Cambridge, MA: MIT Press.

Miller, Clark. 2017. It's Not a War on Science. *Issues in Science and Technology* (Spring): 26–30.

Miller, Clark, and Paul N. Edwards, eds. 2001. *Changing the Atmosphere: Expert Knowledge and Environmental Governance*. Cambridge, MA: MIT Press.

Mirowski, Philip. 2009. Postface: Defining Neoliberalism. In *The Road From Mont Pelerin: The Making of a Neoliberal Thought Collective*, ed. Philip Mirowski and Dieter Plehwe, 417–456. Cambridge, MA: Harvard University Press.

Mirowski, Philip. 2011. *Science-Mart: Privatizing American Science*. Cambridge, MA: Harvard University Press.

Mirowski, Philip, and Esther-Miriam Sent. 2008. The Commercialization of Science and the Response of STS. In *The Handbook of Science and Technology Studies*, ed. Edward Hackett, Olga Amsterdamska, Michael Lynch, and Judy Wajcman, 635–689. Cambridge, MA: MIT Press.

Mitchell, Timothy. 2011. *Carbon Democracy: Political Power in the Age of Oil*. London: Verso.

Monckeberg, Monica. 2007. *El negocio de las universidades en Chile*. Santiago: Random House.

Montecino, Vivian, and Julieta Orlando, eds. 2015. *Ciencias ecológicas 1983–2013: Treinta años de investigaciones chilenas*. Santiago: Editorial Universitaria.

Moon, Suzanne. 2007. *Technology and Ethical Idealism*. Leiden: CNWS Publications.

Moore, Kelly, Daniel Lee Kleinman, David Hess, and Scott Frickel. 2011. Science and Neoliberal Globalization: A Political Sociological Approach. *Theory and Society* 40 (5): 505–532.

Mosciatti, Tomás 2011. Las razones para rechazar el proyecto HidroAysén. CNN Chile, May 9.

Moulian, Tomás. 2002. *Chile actual: Anatomía de un mito.* Santiago: LOM Ediciones.

Moya, Verónica, and Nadia Cabello. 2011. Funcionarios que evaluarán HidroAysén reciben presión de opositores al proyecto. *El Mercurio*, April 21.

Mulsow, Sandor, and Mariano Grandjean. 2006. Incompatibility of Sulphate Compounds and Soluble Bicarbonate Salts in the Rio Cruces Waters: An Answer to the Disappearance of Egeria Densa and Black-Necked Swans in a RAMSAR Sanctuary. *Ethics in Science and Environmental Politics* 6:5–11.

Muñoz-Erickson, Tisha A., Clark A. Miller, and Thaddeus R. Miller. 2017. How Cities Think: Knowledge Co-Production for Urban Sustainability and Resilience. *Forests* 8 (6): 203–220.

Nahuelpan, Héctor. 2016. Micropolíliticas mapuche contra el despojo en el Chile neoliberal: La disputa por el *lafkenmapu* (territorio costero) en Mehuín. *Izquierdas* 30 (October): 89–123.

Najam, Adil. 2005. The View from the South: Developing Countries in Global Environmental Politics. In *The Global Environment: Institutions, Law, and Policy*, ed. Regina S. Axelrod, David L. Downie, and Norman J. Vig, 239–258. Washington, DC: CQ Press.

Nardini, Andrea, Hernán Blanco, and Carmen Senior. 1997. Why Didn't EIA Work in the Chilean Project Canal Laja-Diguillín? *EIA Review* 17 (1): 53–63.

Navarrete, Jorge. 2011. Es la sociedad civil, estúpido! *La Tercera*, May 29.

Nef, Jorge. 1995. Environmental Policy and Politics in Chile: A Latin American Case Study. In *Environmental Policies in the Third World: A Comparative Analysis*, ed. O. P. Dwivedi and Dhirendra K. Vajpeyi, 145–174. Westport, CT: Greenwood Press.

Nelkin, Dorothy. 1977. Scientists and Professional Responsibility: The Experience of American Ecological Society. *Social Studies of Science* 7 (1): 75–95.

Nespolo, Roberto F., Paulina Artacho, Claudio Verdugo, and Luis E. Castañeda. 2008. Short-Term Thermoregulatory Adjustments in a South American Anseriform, the Black-Necked Swan (Cygnus Melanocoryphus). *Comparative Biochemistry and Physiology Part A: Molecular & Integrative Physiology* 150 (3): 366–368.

Niklitschek, Edwin J., Doris Soto, Alejandra Lafon, Carlos Molinet, and Pamela Toledo. 2013. Southward Expansion of the Chilean Salmon Industry in the Patagonian Fjords: Main Environmental Challenges. *Reviews in Aquaculture* 5 (3): 172–195.

Nixon, Rob. 2013. *Slow Violence and the Environmentalism of the Poor.* Cambridge, MA: Harvard University Press.

Norambuena, Cecilia, and Francisco Bozinovic. 2009. Health and Nutritional Status of a Perturbed Black-Necked Swan (Cygnus Melanocoryphus) Population: Diet Quality. *Journal of Zoo and Wildlife Medicine* 40 (4): 607–616.

OECD. 2005. *Environmental Performance Reviews: Chile*. Paris: Organization for Economic Cooperation and Development.

OECD. 2009. *La educación superior en Chile*. Paris: Organization for Economic Cooperation and Development.

OLCA. 2007. Cuestionan a auspiciadores de Encuentro de Ecólogos. *Comunicaciones OLCA*, September 30. Accessed November 9, 2017, http://www.olca.cl/oca/chile /region03/pascualama279.htm.

Oreskes, Naomi, and Erik Conway. 2011. *Merchants of Doubt: How a Handful of Scientists Obscured the Truth on Issues from Tobacco Smoke to Global Warming*. New York: Bloomsbury Press.

Ortolano, Leonard, Bryan Jenkins, and Ramon Abracosa. 1987. Speculations on When and Why EIA Is Effective. *EIA Review* 7 (4): 285–292.

Ossa, Manuel. 2001. Un Valle se Muere. *Pastoral Popular*, September–October, n.p.

Ottinger, Gwen. 2013. *Refining Expertise: How Responsible Engineers Subvert Environmental Justice*. New York: New York University Press.

Owens, Susan. 2015. *Knowledge, Policy, and Expertise: The UK Royal Commission on Environmental Pollution, 1970–2011*. Oxford: Oxford University Press.

Owens, Susan, and Richard Cowell. 2002. *Land and Limits: Interpreting Sustainability in the Planning Process*. London: Routledge.

Owens, Susan, Tim Rayner, and Olivia Bina. 2004. New Agendas for Appraisal: Reflections on Theory, Practice, and Research. *Environment and Planning A* 36 (11): 1943–1959.

Oyarzo, Mariela. 2014. El imaginario social construido por la prensa chilena sobre la contaminación del Río Cruces en Valdivia. PhD diss., Autonomous University of Barcelona.

Palma, Alvaro, Marcelo Silva, Carlos Munoz, Carolina Cartes, and F. M. Jaksic. 2008. Effect of Prolonged Exposition to Pulp Mill Effluents on the Invasive Aquatic Plant Egeria Densa and Other Primary Producers: A Mesocosm Approach. *Environmental Toxicology and Chemistry* 27 (2): 387–396.

Pielke, Roger. 2007. *The Honest Broker: Making Sense of Science in Policy and Politics*. Cambridge: University of Cambridge Press.

Pinochet, Dante, Carlos Ramirez, Roberto MacDonald, and L. Riedel. 2004. Concentraciones de elementos minerales en Egeria Densa Planch. Colectada en el Santuario de la Naturaleza Carlos Anwandter, Valdivia, Chile. *Agro Sur* 32 (2): 80–86.

Pizarro, Carolina. 2014. Hernán Salazar: "Un Comité de Ministros, por su naturaleza, decide políticamente." *La Tercera*, June 14.

Plehwe, Dieter. 2009. Introduction to *The Road from Mont Pelerin*, ed. Philip Mirowski and Dieter Plehwe, 1–44. Cambridge, MA: Harvard University Press.

Pope, Jenny, Alan Bond, Angus Morrison-Saunders, and Francois Retief. 2013. Advancing the Theory and Practice of Impact Assessment: Setting the Research Agenda. *EIA Review* 41 (July): 1–9.

Porter, Ted. 1995. *Trust in Numbers: The Pursuit of Objectivity in Public Life*. Princeton, NJ: Princeton University Press.

Prakash, Gyan. 1999. *Another Reason: Science and the Imagination in Modern India*. Princeton, NJ: Princeton University Press.

Price, Don K. 1965. *The Scientific Estate*. Cambridge, MA: Harvard University Press.

Prieto, Manuel, and Carl Bauer. 2012. Hydroelectric Power Generation in Chile: An Institutional Critique of the Neutrality of Market Mechanisms. *Water International* 37 (2): 131–146.

Prof. Eugenio Figueroa es el nuevo Director Ejecutivo del CENMA. 2003. University of Chile, April 2. http://www.uchile.cl/acerca/rectoria/noticias/02abril2003.html.

Propuesta Comisión Sindical Parlamentaria para la Reforma al Sistema de Evaluación de Impacto Ambiental. 2016. FIMA. Accessed August 26, 2016, http://www.chilesustentable.net/propuestas-comision-sindical-ciudadana-parlamentaria-para-la-reforma-al-sistema-de-evaluacion-de-impacto-ambiental/.

Quesada, Fernando. 2013. Private Foreign Aid and the Contest for Academic Autonomy: The Rockefeller Foundation at the University of Chile. In *The Politics of Academic Autonomy in Latin America*, ed. Fernanda Beigel, 157–172. Surrey, UK: Ashgate Press.

Quiroga, Rayén, and Sara Larraín. 1994. *El tigre sin selva: Consecuencias ambientales de la transformación económica de Chile, 1974–1993*. Santiago: Instituto de Ecología Política.

Rabatel, A., H. Castebrunet, V. Favier, L. Nicholson, and C. Kinnard. 2011. Glacier Changes in the Pascua-Lama Region, Chilean Andes (29° S): Recent Mass Balance and 50 Yr Surface Area Variations. *Cryosphere* 5 (4): 1029–1041.

Ramirez, Carlos, Eewin Carrasco, Silvana Mariani, and Nicolas Palacios. 2006. La desaparición del luchecillo (Egeria Densa) del Santuario del Río Cruces (Valdivia, Chile): Una hipótesis plausible. *Ciencia & Trabajo* 8 (20): 79–86.

Ramsar. 2005. *Informe de misión: Santuario Carlos Anwandter (Río Cruces), Chile*. Prepared by Walter di Marzio and Rob McInnes for Conama. Accessed October 30, 2017, http://www.sinia.cl/1292/articles-31988_informe_ramsar.pdf.

Retief, Francois, Alan Bond, Jenny Pope, Angus Morrison-Saunders, and Nicholas King. 2016. Global Megatrends and Their Implications for Environmental Assessment Practice. *EIA Review* 61 (November): 52–60.

Roblero, Maria. 1995. *Promesas del asombro: Hector Croxatto: Un pionero de la ciencia experimental en Chile*. Santiago: Ediciones Universidad Catolica de Chile.

Rodríguez Medina, Leandro. 2014. *Centers and Peripheries in Knowledge Production*. New York: Routledge.

Ross, Jen. 2005. Glaciers under Threat by Mining in Chile. *NACLA Report on the Americas* 39 (November–December): 16–19.

Ross Schneider, Ben. 2004. *Business Politics and the State in Twentieth-Century Latin America*. Cambridge: Cambridge University Press.

Rozema, Jaap G., Alan J. Bond, Matthew Cashmore, and Jason Chilvers. 2012. An Investigation of Environmental and Sustainability Discourses Associated with the Substantive Purposes of Environmental Assessment. *EIA Review* 33 (1): 80–90.

Rueschemeyer, Dietrich, and Theda Skocpol. 1996. *States, Social Knowledge, and the Origins of Modern Social Policies*. Princeton, NJ: Princeton University Press.

Ruthenberg, Ina-Marlene. 2001. *A Decade of Environmental Management in Chile*. Washington, DC: World Bank Group.

Sábato, Jorge. 2004. *Ensayos en Campera*. Buenos Aires: Universidad Nacional de Quilmes Editorial. First published 1979.

Sagar, Ambuj, and Stacy VanDeveer. 2005. Capacity Development for the Environment: Broadening the Scope. *Global Environmental Politics* 5 (3): 14–22.

Salas, María José. 2012. Las zonas de sacrificio. *Revista Capital*, October 28.

Salmon Crisis, Red Tide to Double Unemployment in Southern Chilean Region. 2016. *Undercurrent News*, May 30.

Salmon Roundtable. 2009. *Informe Final: Comisión Asesora Ministerial, denominada "Grupo de Tareas del Salmón."* On file with author.

Sannazzaro, Jorgelina. 2014. Citizen Cartography, Strategies of Resistance to Established Knowledge, and Collective Forms of Knowledge Building. *Public Understanding of Science* 25 (3): 346–360.

Sarewitz, Daniel. 2004. How Science Makes Environmental Controversies Worse. *Environmental Science and Policy* 7 (5): 385–403.

Sayre, Nathan. 2008. The Genesis, History, and Limits of Carrying Capacity. *Nature and Society* 98 (1): 120–134.

Schaeffer, Colombina, and Mattijs Smits. 2015. From Matters of Fact to Places of Concern? Energy, Environmental Movements, and Place-Making in Chile and Thailand. *Geoforum* 65 (October): 146–157.

Schell, Patience. 2013. *The Sociable Sciences: Darwin and His Contemporaries in Chile*. New York: Palgrave Macmillan.

Schurman, Rachel. 2004. Shuckers, Sorters, Headers, and Gutters: Labor in the Fisheries Sector. In *Victims of the Chilean Miracle*, ed. Peter Winn, 298–336. Durham, NC: Duke University Press.

Schwartzman, Simon. 1987. Peripheral Science. *Social Studies of Science* 17 (3): 1–6.

Schwartzman, Simon, ed. 2008. *Universidad y desarrollo en Latinoamérica: Experiencias exitosas de centros de investigación.* Bogotá: IESALC-UNESCO.

Scott, James C. 1998. *Seeing Like a State: How Certain Schemes to Improve the Human Condition Have Failed.* New Haven, CT: Yale University Press.

Sepúlveda, Claudia. 2016. Swans, Ecological Struggles, and Ontological Fractures: A Posthumanist Account of the Rio Cruces Disaster in Valdivia, Chile. PhD diss., University of British Columbia.

Sepúlveda, Claudia, and Pablo Villarroel. 2012. Swans, Conflicts, and Resonance: Local Movements and the Reform of Chilean Environmental Institutions. *Latin American Perspectives* 39 (4): 181–200.

Seremi acusó presiones ante votación de HidroAysén: Me siento en una película. 2011. *Cooperativa.cl*, April 21.

Sernapesca. 2005. *Diagnóstico ambiental de la acuicultura chilena en función de estándares establecidos en el Reglamento Ambiental para la Acuicultura.* On file with author.

Sernapesca. 2008. *Balance de la situación sanitaria de la Anemia Infecciosa del Salmón en Chile, de julio del 2007 a julio del 2008.* On file with author.

Sernapesca. 2009. *Informe final propuesta: Evaluación de un sistema de control de la bioseguridad para monitorear en línea factores de riesgo en la propagación de enfermedades en la futura actividad salmonera.* On file with author.

Sernapesca. 2010. *Programa sanitario específico de vigilancia y control de la Anemia Infecciosa del Salmón (PSEC-ISA). Período: noviembre 2008–noviembre 2010.* On file with author.

Shrader-Frechette, Kristin. 2014. *Tainted: How Philosophy of Science Can Expose Bad Science.* Oxford: Oxford University Press.

Shrum, Wesley, and Yehouda Shenhav. 1995. Science and Technology in Less Developed Countries. In *The Handbook of Science and Technology Studies*, ed. Sheila Jasanoff, Gerald Markle, James Peterson, and Trevor Pinch, 627–651. Cambridge, MA: MIT Press.

Silva, Eduardo. 1994. Contemporary Environmental Politics in Chile: The Struggle over the Comprehensive Law. *Industrial and Environmental Crisis* 8 (4): 323–343.

Silva, Eduardo. 1996. Democracy, Market Economics, and Environmental Policy in Chile. *Journal of Interamerican Studies and World Affairs* 38 (4): 1–33.

Silva, Eduardo. 1997. Chile. In *National Environmental Policies: A Comparative Study of Capacity Building*, ed. Martin Jänicke and Helmut Weidner, 213–235. Berlin: Springer.

Silva, Patricio. 2009. *In the Name of Reason: Technocrats and Politics in Chile*. University Park: Penn State University Press.

Skewes, Juan Carlos. 2004. Conocimiento científico y conocimiento local: Lo que las universidades no saben acerca de lo que actores locales saben. *Cinta de Moebio*. March 19.

Skewes, Juan Carlos, and Debbie Guerra. 2004. The Defense of Maiquillahue Bay: Knowledge, Faith, and Identity in an Environmental Conflict. *Ethnology* 43:217–231.

Slaughter, Sheila, and Gary Rhoades. 2004. *Academic Capitalism and the New Economy: Markets, State, and Higher Education*. Baltimore: Johns Hopkins University Press.

Soluri, John. 2011. Something Fishy: Chile's Blue Revolution, Commodity Diseases, and the Problem of Sustainability. *Latin American Research Review* 46:55–81.

Soto, Hector. 2011. La Ultima Utopia. *La Tercera*, May 15.

Soto Laveaga, Gabriela. 2009. *Jungle Laboratories: Mexican Peasants, National Projects, and the Making of the Pill*. Durham, NC: Duke University Press.

Stein, Ernesto, Mariano Tommasi, Koldo Echebarría, and Eduardo Lora. 2005. *The Politics of Policies: Economic and Social Progress in Latin America*. Washington, DC: Inter-American Development Bank.

Steinberg, Paul. 2001. *Environmental Leadership in Developing Countries: Transnational Relations and Biodiversity Policy in Costa Rica and Bolivia*. Cambridge, MA: MIT Press.

Steinberg, Paul, and Stacy D. VanDeveer. 2012. Comparative Environmental Politics in a Global World. In *Comparative Environmental Politics: Theory, Practice, Prospects*, ed. Paul Steinberg and Stacy VanDeveer, 3–28. Cambridge, MA: MIT Press.

Storey, William K. 1997. *Science and Power in Colonial Mauritius*. Rochester, NY: University of Rochester Press.

Subpesca. 2002. *Fundamentos para la elaboración de la resolución acompañante del Reglamento Ambiental para la Acuicultura, D.S. No. 320/2001 (MINECON). Informe Técnico Ambiental No. 11*. Prepared by Alex W. Brown, Juan Antonio Manríquez, and Paula Moreno. On file with author.

Subpesca. 2006. *Informe ambiental de la acuicultura*. On file with author.

Subpesca. 2008a. *Informe ambiental de la acuicultura, 2005–2006*. On file with author.

Subpesca. 2008b. *Notas del Taller: Análisis de los parámetros de evaluación ambiental y límites de aceptabilidad en el ámbito del Reglamento Ambiental para la Acuicultura (RAMA)*. On file with author.

Sumathi, Suresh, and Yung-Tse Hung. 2005. Treatment of Pulp and Paper Mill Wastes. In *Waste Treatment in the Process Industries*, ed. Lawrence K. Wang, Yung-Tse Lung, Howard H. Lo, and Constantine Yapijakis, 453–498. Boca Raton: CRC Press.

SustenTank. 2011. *Evaluación y diseño definitivo del Centro de Referencia Ambiental*. Prepared for CENMA, June 20. On file with author.

Taillant, Jorge D. 2015. *Glaciers: The Politics of Ice*. Oxford: Oxford University Press.

Taranger, Geir Lasse, Orjan Karlsen, Raymond J. Bannister, Kevin A. Glover, Vivien Husa, Egil Karsbakk, Bjorn O. Kvamme, et al. 2015. Risk Assessment of the Environmental Impact of Norwegian Atlantic Salmon Farming. *ICES Journal of Marine Science* 72 (3): 1–25.

Taylor, Matthew E. 1998. Economic Development and the Environment in Chile. *The Journal of Environment and Development* 7 (4): 422–436.

Tecklin, David, Carl Bauer, and Manuel Prieto. 2011. Making Environmental Law for the Market: The Emergence, Character, and Implications of Chile's Environmental Regime. *Environmental Politics* 20 (6): 879–898.

Thompson, Charis. 2004. Co-Producing CITES and the African Elephant. In *States of Knowledge: The Co-Production of Science and Social Order*, ed. Sheila Jasanoff, 67–86. London: Routledge.

Tilley, Helen. 2011. *Africa as a Living Laboratory*. Chicago: University of Chicago Press.

Tinsman, Heidi. 2014. *Buying into the Regime: Grapes and Consumption in Cold War Chile and the United States*. Durham, NC: Duke University Press.

Tironi, Eugenio. 2011. *Abierta: Gestión de controversias y justificaciones*. Santiago: Uqbar.

Tironi, Manuel, and Javiera Barandiarán. 2014. Neoliberalism as Political Technology: Expertise, Energy, and Democracy in Chile. In *Beyond Imported Magic: Essays on Science, Technology, and Society in Latin America*, ed. Eden Medina, Ivan Costa Marquez, and Christina Holmes, 305–330. Cambridge, MA: MIT Press.

Una estrategia condenable y peligrosa. 2011. *La Tercera*, April 24.

University of Chile. 2008. HidroAysén Geological Baseline Reports. Prepared by the Department of Geology, February 2011.

Ureta, Sebastián. 2015. *Assembling Policy: Transantiago, Human Devices, and the Dream of a World-Class Society*. Cambridge, MA: MIT Press.

Urkidi, Leire. 2010. A Glocal Environmental Movement against Gold Mining: Pascua-Lama in Chile. *Ecological Economics* 70 (2): 219–227.

Urkidi, Leire, and Mariana Walter. 2011. Dimensions of Environmental Justice in Anti-Gold Mining Movements in Latin America. *Geoforum* 42 (6): 683–695.

Vaccarezza, Leonardo. 2006. Autonomia universitaria, reformas y transformacion social. In *Universidad e investigacion científica: Convergencias y tendencias*, ed. Hebe Vessuri. Buenos Aires: CLACSO.

Valdés, Juan Gabriel. 1995. *Pinochet's Economists: The Chicago School in Chile*. Cambridge: Cambridge University Press.

VanDeveer, Stacy. 2004. Ordering Environments: Organizing Knowledge and Regions in European International Environmental Cooperation. In *Earthly Politics: Local and Global in Environmental Governance*, ed. Sheila Jasanoff and Marybeth Long Martello, 309–334. Cambridge, MA: MIT Press.

VanDeveer, Stacy, and Geoffrey Dabelko. 2001. It's Capacity, Stupid: International Assistance and National Implementation. *Global Environmental Politics* 1 (2): 18–29.

Venezian, Eduardo. 1987. *Chile and the CGIAR Centers: A Study of Their Collaboration in Agricultural Research*. Washington, DC: World Bank Group.

Vessuri, Hebe. 1987. The Social Study of Science in Latin America. *Social Studies of Science* 17 (3): 519–554.

Vessuri, Hebe. 1988. The Universities, Scientific Research, and the National Interest in Latin America. *Minerva* 24 (1): 1–38.

Vessuri, Hebe. 1994. Foreign Scientists, the Rockefeller Foundation, and the Origins of Agricultural Science in Venezuela. *Minerva* 32 (3): 267–296.

Vessuri, Hebe. 2007. *"O inventamos o erramos": La ciencia como idea-fuerza en América Latina*. Bernal, Buenos Aires: Universidad Nacional de Quilmes Editorial.

Vessuri, Hebe, Jean Claude Guedon, and Ana María Cetto. 2014. Excellence or Quality? Impact of the Current Competition Regime on Science and Scientific Publishing in Latin America and Its Implications for Development. *Current Sociology* 62 (5): 647–665.

Villalobos, Sergio. 1990. *Historia de la ingenieria en Chile. Santiago*. Santiago: Ediciones Pedagogicas Chilenas.

Waterton, Claire, and Brian Wynne. 2004. Knowledge and Political Order in the European Environment Agency. In *States of Knowledge: The Co-Production of Science and Social Order*, ed. Sheila Jasanoff, 87–108. London: Routledge.

Weber, Anila. 2011. "Chile a la Luz" es el novedoso arte visual que se proyecta sobre las aguas de un río de la capital chilena. *La Gran Epoca*, January 21.

Weber, Max. 1946. Bureaucracy. In *From Max Weber: Essays in Sociology*, ed. and trans. H. H. Gerth and C. Wright Mills, 196–244. New York: Oxford University Press.

Wilkening, Kenneth E. 2004. *Acid Rain Science and Politics in Japan*. Cambridge, MA: MIT Press.

Winn, Peter, ed. 2004. *Victims of the Chilean Miracle: Workers and Neoliberalism in the Pinochet Era, 1973–2002*. Durham, NC: Duke University Press.

World Bank. 1989. *Operational Directive on Environmental Assessment*. Washington, DC: World Bank Group.

World Bank. 1992. *The World Bank and the Environment: Annual Report: 1992*. Washington, DC: World Bank Group.

Wormald, Guillermo, and Daniel Brieba. 2012. Institutional Change and Development in Chilean Market Society. In *Institutions Count: Their Role and Significance in Latin American Development*, ed. Alejandro Portes and Lori D. Smith, 60–84. Berkeley: University of California Press.

WWF. 2005. *Informe de observaciones y recomendaciones: Misión internacional de evaluación de WWF ante la controversia del Santuario de la Naturaleza y sitio Ramsar Carlos Anwandter y la planta de celulosa Valdivia de Celco*. Prepared by the World Wildlife Fund. Accessed October 30, 2017, http://www.wwf.cl/?199946/Informe-de-observaciones-y -recomendaciones-de-WWF-por-controversia-Santuario-de-la-Naturaleza-y-Celulosa -Arauco.

Yañez, Nancy, and Sarah Rae. 2006. The Valley of Gold. *Cultural Survival Quarterly* 30 (4): 41–45.

Yearley, Steven. 1996. Nature's Advocates: Putting Science to Work in Environmental Organisations. In *Misunderstanding Science? The Public Reconstruction of Science and Technology*, ed. Alan Irwin and Brian Wynne, 172–190. Cambridge: Cambridge University Press.

Zabala, Juan Pablo. 2010. *La enfermedad del Chagas en la Argentina: Investigación científica, problemas sociales, y políticas sanitarias*. Buenos Aires: Universidad Nacional de Quilmes Editorial.

Zaror, Claudio. 2005. *Informe final: Apoyo al análisis de fuentes de emisión de gran magnitud y su influencia sobre los ecosistemas de la subcuenca del Río Cruces*. Prepared for Conama. Accessed October 30, 2017, http://www.sinia.cl/1292/printer-33579 .html.

Index

Urban and Industrial Environments

Series editor: Robert Gottlieb, Henry R. Luce Professor of Urban and Environmental Policy, Occidental College

Julie Sze, *Noxious New York: The Racial Politics of Urban Health and Environmental Justice*

Robert D. Bullard, ed., *Growing Smarter: Achieving Livable Communities, Environmental Justice, and Regional Equity*

Ann Rappaport and Sarah Hammond Creighton, *Degrees That Matter: Climate Change and the University*

Michael Egan, *Barry Commoner and the Science of Survival: The Remaking of American Environmentalism*

David J. Hess, *Alternative Pathways in Science and Industry: Activism, Innovation, and the Environment in an Era of Globalization*

Peter F. Cannavò, *The Working Landscape: Founding, Preservation, and the Politics of Place*

Paul Stanton Kibel, ed., *Rivertown: Rethinking Urban Rivers*

Kevin P. Gallagher and Lyuba Zarsky, *The Enclave Economy: Foreign Investment and Sustainable Development in Mexico's Silicon Valley*

David N. Pellow, *Resisting Global Toxics: Transnational Movements for Environmental Justice*

Robert Gottlieb, *Reinventing Los Angeles: Nature and Community in the Global City*

David V. Carruthers, ed., *Environmental Justice in Latin America: Problems, Promise, and Practice*

Tom Angotti, *New York for Sale: Community Planning Confronts Global Real Estate*

Paloma Pavel, ed., *Breakthrough Communities: Sustainability and Justice in the Next American Metropolis*

Anastasia Loukaitou-Sideris and Renia Ehrenfeucht, *Sidewalks: Conflict and Negotiation over Public Space*

David J. Hess, *Localist Movements in a Global Economy: Sustainability, Justice, and Urban Development in the United States*

Julian Agyeman and Yelena Ogneva-Himmelberger, eds., *Environmental Justice and Sustainability in the Former Soviet Union*

Jason Corburn, *Toward the Healthy City: People, Places, and the Politics of Urban Planning*

JoAnn Carmin and Julian Agyeman, eds., *Environmental Inequalities Beyond Borders: Local Perspectives on Global Injustices*

Louise Mozingo, *Pastoral Capitalism: A History of Suburban Corporate Landscapes*

Gwen Ottinger and Benjamin Cohen, eds., *Technoscience and Environmental Justice: Expert Cultures in a Grassroots Movement*

Samantha MacBride, *Recycling Reconsidered: The Present Failure and Future Promise of Environmental Action in the United States*

Andrew Karvonen, *Politics of Urban Runoff: Nature, Technology, and the Sustainable City*

Daniel Schneider, *Hybrid Nature: Sewage Treatment and the Contradictions of the Industrial Ecosystem*

Catherine Tumber, *Small, Gritty, and Green: The Promise of America's Smaller Industrial Cities in a Low-Carbon World*

Sam Bass Warner and Andrew H. Whittemore, *American Urban Form: A Representative History*

John Pucher and Ralph Buehler, eds., *City Cycling*

Stephanie Foote and Elizabeth Mazzolini, eds., *Histories of the Dustheap: Material Cultures, Social Justice*

David J. Hess, *Good Green Jobs in a Global Economy: Making and Keeping New Industries in the United States*

Joseph F. C. DiMento and Clifford Ellis, *Changing Lanes: Visions and Histories of Urban Freeways*

Joanna Robinson, *Contested Water: The Struggle Against Water Privatization in the United States and Canada*

William B. Meyer, *The Environmental Advantages of Cities: Countering Commonsense Antiurbanism*

Rebecca L. Henn and Andrew J. Hoffman, eds., *Constructing Green: The Social Structures of Sustainability*

Peggy F. Barlett and Geoffrey W. Chase, eds., *Sustainability in Higher Education: Stories and Strategies for Transformation*

Isabelle Anguelovski, *Neighborhood as Refuge: Community Reconstruction, Place Remaking, and Environmental Justice in the City*

Kelly Sims Gallagher, *The Globalization of Clean Energy Technology: Lessons from China*

Vinit Mukhija and Anastasia Loukaitou-Sideris, eds., *The Informal American City: Beyond Taco Trucks and Day Labor*

Roxanne Warren, *Rail and the City: Shrinking Our Carbon Footprint While Reimagining Urban Space*

Marianne E. Krasny and Keith G. Tidball, *Civic Ecology: Adaptation and Transformation from the Ground Up*

Erik Swyngedouw, *Liquid Power: Contested Hydro-Modernities in Twentieth-Century Spain*

Ken Geiser, *Chemicals without Harm: Policies for a Sustainable World*

Duncan McLaren and Julian Agyeman, *Sharing Cities: A Case for Truly Smart and Sustainable Cities*

Jessica Smartt Gullion, *Fracking the Neighborhood: Reluctant Activists and Natural Gas Drilling*

Nicholas A. Phelps, *Sequel to Suburbia: Glimpses of America's Post-Suburban Future*

Shannon Elizabeth Bell, *Fighting King Coal: The Challenges to Micromobilization in Central Appalachia*

Theresa Enright, *The Making of Grand Paris: Metropolitan Urbanism in the Twenty-First Century*

Robert Gottlieb and Simon Ng, *Global Cities: Urban Environments in Los Angeles, Hong Kong, and China*

Anna Lora-Wainwright, *Resigned Activism: Living with Pollution in Rural China*

Scott L. Cummings, *Blue and Green: The Drive for Justice at America's Port*

David Bissell, *Transit Life: Cities, Commuting, and the Politics of Everyday Mobilities*

Javiera Barandiarán, *Science and Environment in Chile: The Politics of Expert Advice in a Neoliberal Democracy*